DECADES OF DIOXIN

DECADES OF DIOXIN

Limelight On a Molecule

WARREN B. CRUMMETT

Front cover: Herr Doktor Professor Otto Hutzinger welcomes the author to Mitwitz
Castle, Bavaria, Germany

.

This book was printed in the United States of America.

To order additional copies of this book, contact:
Xlibris Corporation
1-888-795-4274
www.Xlibris.com
Orders@Xlibris.com
16064-CRUM

TO

Wife Elizabeth and sons Allan and Daniel
who understood my concerns.

Secretary Connie March, who became a partner
in producing this book.

All colleagues in "dioxin" matters, especially the "Dow School".

Members of the "Dow School of Analytical Measurement", who
produced extraordinarily fine data beyond refutation.

CONTENTS

PREFACE

"Nothing has really happened until it has been recorded",
Virginia Wolf (1882-1941).

I am now an octogenarian and, thinking about life, I am haunted by the notion that, when the natural forces designed to assure my demise succeed, a library will be lost forever. The library consists of memories and documents relating to the trails I've followed and the new ones blazed. Some are unmarked, remote and less traveled. Others are well marked and worn. The trails touch on the spiritual, intellectual, physical, social, political and scientific. Many of the trails were not deliberately chosen but arose from the unique opportunities presented by the adventures of day-to-day living.

Of the many trails, the one best documented is that dealing with a molecule called "dioxin", and that is what this book is about. From 1969 to 1988, "dioxin" considerations consumed about half of my professional work time and the trail almost became a highway. This book describes some unusual experiences along the way, together with commentary.

Some may say that this is an account of myself by myself, for myself, being myself in spite of myself, on account of the muster of others being themselves in creating, devising, and organizing great technical and scientific schemes to find, characterize, and control a molecule called 2,3,7,8-tetrachloro-dibenzo-p-dioxin. However, since it is also about fear, inherent in all of us when faced with the unknown, it is also about you being yourself in spite of yourself. The fear enhanced by the portrayal of this molecule leads to the possibility that it might really be a mischievous, noisy spirit.

11

As a lad I lived among people, some of whom were superstitious and some persons were identified as witches and witch doctors. At that time I read many fairy tales and let my imagination run wild. But I never could have imagined a real world scenario in which an inanimate molecule was the ogre, analytical chemists were knights in shining armor, bureaucrats were the king's men, pseudo scientists were fear mongers, and industrial chemists were witches and their management warlocks.

It happened! It really happened! And I was in the eye of the resulting furor! Moreover it occurred among people whose formal education was well documented and who projected themselves as scientifically literate and rational.

Although not written as a fairy tale, this book will show the grip fear still holds in the minds of people.

And so this book is like a diary whose subject is a molecule. The molecule became notorious because it caused great fear in people as evidenced by their doing strange things. Meanwhile the scientists, of whom I am one, had their integrity questioned in unprecedented ways, as they performed extraordinary experiments in a great effort to learn the truth.

Prof. Christopher Rappe views the Canadian mountains.

INTRODUCTION

"Without adventure, civilization is in full decay",
Alfred North Whitehead.
"Science is the search for truth – it is not a game in which one
tries to beat his opponent, to do harm to others",
Linus Carl Pauling.

In the second half of the twentieth century a little known molecule dramatically emerged on the world scene and quickly received unprecedented coverage by the press, especially in the United States. This undesired trace contaminant was found in some highly useful chemical products and identified as 2,3,7,8-tetrachlorodibenzo-p-dioxin. Quickly named "dioxin" and given the acronym TCDD, the press made a valiant effort to present a full description of this molecule, talking with almost everyone except industrial scientists, who knew it best. The result was a strange mixture of fact and fancy. The fancy part appeared to attribute great powers of death to this molecule and thus created great fear among the "enlightened". Observing this, some, including myself, jokingly wondered whether "dioxin" was really a molecule or an evil spirit.

Early in the century's eighth decade, significant forces started to impact each other. First, the fury of the press developed into a full-blown brouhaha. Second, scientists all over the world were empowered to independently conduct experiments necessary to determine the scientific facts and so interpret the behavior of this molecule and its relatives. Hopefully, these two movements would converge and bring a consensus, which represented the truth.

"Dioxin" was reluctant to reveal its secrets. But the unprecedented power of the newly developed scientific methods produced convincing results. Eventually we learned enough to understand its ancestry, its behavior and its demise. Thus we hoped to totally destroy it. We found that, like other molecules, "dioxin" behaves in totally predictable and reproducible ways. Unlike other molecules, however, "dioxin" has a special propensity to scare people. (Mention its name and

Dioxin scientists gather about Dr. Hans Rudulph Buser
of Switzerland near Ottawa, Canada.

people succumb to notions of fear.) It also appears to have the power to cause communicators to mix fact and fancy. (Even we scientists sometimes fail to stick to the facts.)

As the Technical Manager of Dow's Analytical Laboratories and an active member of several scientific societies, I was in an excellent position to work cooperatively with all scientists and to lead the effort to assure the integrity of analytical data. But first the analytical scientists in the Dow laboratory had to build a world-class reputation which properly reflected their extraordinary competence. To assure this we needed to publish all confirmed information in peer-reviewed scientific journals. Since, in the beginning, our efforts to publish were resisted, I knew the search for the truth about "dioxin" would be long and difficult. It was a unique challenge for industrial analytical chemists. But eventually we reached our goal and I became known by some as an advocate for science.

Being a part of an informal international network of distinguished scientists devoted to producing comparable data, which would define the true behavior of "dioxin" felt great! The collaboration among government, academic and industrial scientists was extraordinary. As the work progressed, some of my concerns were quelled, some remained, and new ones surfaced and were enhanced. These concerns include: the toxicity of "dioxin", the integrity of data, the extrapolation of data, the integrity of scientists, the integrity of regulatory agencies, the consciences of industrial complexes, the scientific literacy of the populace as a whole, and the inability of the media to distinguish between sound and flawed science. I will show how these concerns ebbed, and flowed by relating actual personal stories in which some behavior seemed illogical, irrational, hysterical, whimsical, or hilarious. I also reveal what I have learned about some of these concerns by short essays. In doing so I must confess that my professional life was inextricably entangled in the "dioxin" investigations. Reaching this point had been a great adventure filled with all types of human emotion and logic: from insinuation to scientific proof, disgust to amusement, from falderal to fact, etc. For 18 years I had several such fascinating experiences every work week. It was highly challenging not only to my problem-solving skills, but to my emotions as well. In short, it was a beautiful life!

I started to write some of these surprising and remarkable experiences together with various lessons as I neared retirement. On learning of the first draft, publishers correctly observed that the resulting manuscript would contain no sensational matter and so was not saleable. They urged me to write a monograph, textbook, documentary, or treatise—truth for scientists only! Believing others may benefit from my unique experience and perspective, I am making the writing available. I do so with humility. I feel as naïve as ever, even though I've been on an extraordinary journey.

Surprisingly, I found that I had likely been "contaminated" with "dioxin" from the start of life and, because of my unique career preparation, was destined to research the behavior of this molecule even though I have a strong aversion to "toxic" substances. Thus "dioxin" was one of life's themes. So I have had to add a little more about myself. My first 17 years were in a pristine environment, the way many people say they want to live now. It was idyllic for a child but misery for teenagers and adults who grubbed a living from the land. So I studied very hard to escape! But I assure you that the value system of that society which has a reverence for law, truth, integrity, freedom, and God is what makes the pristine environment wonderful. The tranquility of the mountains, forests, streams, insects, fauna, flora, etc., served to make our appreciation of these values more acute.

DECADE I
THE PRISTINE
LAND

I AM

"Fair Science frown'd not on his humble birth",
Thomas Cray, 1750.

4 April 1922

I burst from my mother into the waiting arms of the kind, but gruff, sawbones from World War I, named Dr. Carl Stover. He gave me an appropriate swat and I let out a cry (heard throughout my grandfather's house). Then I inhaled deeply, filling my lungs with air tainted with smoke and ash from the fireplace, kerosene lamps, and tallow candles. These soot particles must have contained at least a part per quintillion of "dioxin", a toxic substance yet to be discovered.

Meanwhile my mother had fallen into a deep sleep, a result not only from exhaustion but also from the inhalation of chloroform from a saturated cloth over her face. I too went to sleep, probably for the same reasons. Some modern activists (had they been present) would have shuttered and urged the banning of heat and light from this old farmhouse.

So there I was – an imperfect infant in an imperfect world! However, since I was the seventh generation of the Crummett family to live off this land, my chances for surviving the life expectancy time of 55 years was good. Almost all my ancestors had done so!

Fifty-five years later I was studying the scientific data which led to the "trace chemistries of fire hypothesis" and I was breathing slightly more "dioxin".

Grandpap's old farmhouse.

THE WITCH DOCTOR

"The first and great commandment is: Don't let them scare you",
Elmer Davis, American News Commentator (1890-1958)

October 1931

The alarm came unexpectedly! "A witch is coming!!" Nothing more. The alert Mt. Hall School sentinels, together with almost all pupils, ran and hid behind fences, bushes, large rocks and outhouses. Dead silence ensued and only a few pair of eyes could be detected peering from behind cover. Two of us didn't move at all. I was paralyzed with fear and felt rooted to the spot. Up the dirt road came a very large horse at a very slow pace carrying a large man in a tattered army uniform. He sat slumped in the saddle. He had a gray beard that matched his hair and a black patch over his left eye, an eye said to be really glass. He spoke not a word, not even to his horse, and he did not acknowledge our presence. He simply went straight ahead and disappeared down the other side of the hill.

I remember thinking that a witch doctor, having supernatural powers, should have given his horse more energy. This "witch doctor" appeared to my 9-year-old mind, to be only a handicapped, lonely man living as best he could in a hollow in the Shenandoah Mountains.

My Mt. Hall School classmates had behaved as the chicks of the partridge – blending into the landscape at the alarm of the mother hen. They had been carefully taught by their mothers to avoid contact with strangers – especially those that our society did not understand, including tramps, gypsies, salesmen, and witches.

I later learned that the "witch doctor" was really skilled in the art of husbandry. He would call on farmers and recommend treatments along with incantations, and charge a small fee. Usually the treatment was just based on experience. He seemed to always know

Mt. Hall School.

when animals should be moved to other pasture fields to avoid various infestations. The people were grateful but, when the treatment worked, many were fearful.

BUGS

"The prevailing rate for picking potato bugs was one cent a hundred",

Herbert Hoover, in his Memoirs.

Summer 1932

The nosey woman on the gravel road stopped under a cherry tree to stare at the frail, barefoot lad of 10 years. He was wearing a torn-brim straw hat and blue denim overalls, and was struggling valiantly to lug a 10-quart galvanized pail half full of potato bug larvae across the vegetable garden, through the unpainted wooden picket fence and across the pasture to the ash hopper. There water was boiling in a 20-gallon iron kettle over a wood fire. Arriving there he set the pail down, dipped boiling water from the kettle, nearly spilling some on his legs, and poured the boiling water on the bugs in the pail, instantly scalding them to death. Nauseated by the resulting stench, he stepped back and sat down to rest briefly on a rock. The wind shifted and smoke from the fire burned his nostrils causing him to think that the potato patch was no worse then this. He felt a little faint and wondered if the smoke were poisonous.

He stood up and noticed that the knees of his overalls were worn bare. It occurred to him that the seat of his overalls was probably also as worn and that explained why when he sat down he felt a prick like the stinging of sweat bees. He was embarrassed by the thought that only in winter could he enjoy the luxury of underwear. Worn overalls were to be expected, however, as he had just finished debugging fifteen 150-foot rows of potatoes. When his back had become unbearably tired, he dropped to his knees and finally sat on his but-

Site of potato bugs. View from the cherry tree.

tocks and "rutched" over the rocky soil between the rows. Moving down the row, he used his hands to shake potato bugs and their larvae from the plants into the pail. Most of the adult bugs flew away. Every potato leaf was examined for nits, which were then crushed between two flat stones.

He now dumped the boiled potato bugs on the ground where they waited for bacteria to turn them into soil. Neither bird nor animal would devour them. The pail was rinsed with hot water and hung upside down on the kettle support to dry and air and then he raked ashes over the fire. While attempting to brush the grime from his clothes, he saw his dad in the potato patch inspecting the plants. When they met at the house, his dad said, "Bus, you didn't get all the bugs, but it will have to do. Tomorrow you must rid the bean vines of Mexican bean beetles the same way. After that we will plow and hoe the corn. Then the potato bugs will have to be killed again".

"Bus" (the nickname given to me by my dad) did not respond. He suddenly felt dirty inside and out. A dull pain had developed in the small of his back. He slunk into the house looking for Doan's kidney pills. They usually made him feel clean inside and the pain would go away.

After eating some cherries, the woman moved on.

WEEDS

*"During my early boyhood years on the farm,
weeds spelled misery",
Joseph A. Cocannover, "Weeds – Guardians of the Soil", 1950.*

Summer 1934

Bus dropped his hoe and eagerly headed for the fence. Although he well knew that snakes frequented the fence-row and that bees, hornets, and wasps would likely have nests there, he ran barefoot through the bramble and propelled himself to the top of the six-foot split-rail fence. Some of the rails were very old and rotten and snapped and cracked under his weight, but he ignored that. Selecting a flat spot on the top rail, he turned toward the east and sat resting with the wind at his back. He could look over Shaw's Ridge to the Shenandoah Mountains where Reddish Knob stood high among the other peaks. He could even see the lookout tower outlined against the blue of the sky, solidly anchored on the dark blue of the mountain.

With ten rows of corn already done and forty more waiting for him, he waited for his dad to finish plowing two more rows – enough so that all four hoers could hoe and move down the rows together. It was a back breaking exercise and painful as well because one often stubbed bare toes on field stone or developed blisters and calluses on hands whose owners were too poor to buy gloves. To keep up with the adult hoers, his arms had to move very fast and the hoe had to move accurately. If the hoe glanced off a pebble and cut a corn stalk, a stiff reprimand would be certain.

His mood became as blue as the mountain. Was there ever to be a respite in his struggle with weeds? Or was he trapped like his ancestors in a never-ending battle with weeds, brush and bugs? He saw no clear path out of the situation. To make matters worse, the original homestead had been divided many times. There was no more land to divide. His grandfather had six sons – a number necessary to slay the weeds and bugs and tend the fields and animals. Was there a better way of life across that mountain?

Suddenly he realized that his mouth was parched. He saw a wild grapevine clinging to the fence. He plucked several of its tendrils and chewed them ever so slowly. The tartness of the tendril caused saliva to flow, watering his mouth and causing him to feel better. He would have preferred to run to the mountain spring for a drink, but the nearest one was a quarter of a mile away and his dad would not let him go. If he did make the trip, he would have had to fetch a jug of water for the rest of the hands, and that may have taken much of the afternoon. More likely he would have forgotten to return at all.

His daydreaming was interrupted by, "Time to hoe, boys!" Grandpa's hoe was already destroying fox tail, crab grass, lambs quarter, ragweed, burdock, Queen Anne's lace, sheep sorrel, various briars, etc. Bus grabbed his hoe and swung it with vigor, trying to please. Together they went down the rows, across the field, exchanging stories, telling jokes, singing, or whistling; beating the boredom to death with the voice. There would be more sore feet, more blisters, more aches, but there would be corn to feed the pigs and chickens all winter. There would be corn bread, corn pone, and corn mush on the table. There would be corncob pipes and corncob and corn husk dolls. There would be fodder in the barn. There would also be plenty of corncobs for the outhouse. All this and more from organic farming.

Next spring we would return everything to the fields. There would be horse manure, cow manure, sheep manure, chicken manure – all to be spread by hand.

Unknown to "Bus", academic social scientists were already discussing the great value of leisure time to man, recognizing that most of mankind would never be able to enjoy it so long as weeds and brush had to be removed by hand to cultivate the fields. Chemists and biologists were beginning to explore the possibility of using chemicals to control weeds. Two top scientists, who would become well known to "Bus" much later, were John Davidson and Wendell Mullison of The Dow Chemical Company. Before this would happen, he would first go over the mountains to the east and then back over them to the west and finally north to Michigan.

Above the cornfield, "Grandpap" Crummett mixes
work and fun.

DECADE II
TOIL & TRIALS

FIRE

"There is no fire without some smoke", John Heywood, c.1493 – c.1580.

Autumn 1932

I carefully positioned the splinters and knots of pitch pine just inside the huge iron stove. The seasoned wood was stacked precisely over the kindling. I struck a match and lit the pine. A yellow flame leaped forth together with thick black smoke. The flames got bigger and the smoke filled the stove. The wind outside shifted and suddenly a down draft blew the black smoke into my face, choking me and making me cough. I was starting the fire in Mt. Hall School, a one-room schoolhouse. My grandfather had contracted with the school board to provide the wood and start the fires. He had subcontracted starting the fires to me at five cents per day.

Smoke in the face was a common occurrence when fires were being started. We knew the smoke was toxic, but we didn't know that in addition to the acrid components, it contained at least a part per quintillion of many chlorinated organic compounds, including "dioxin".

The seasoned wood ignited and the draft stabilized and soon the fire was burning vigorously. By the time the other school children (6-20 year olds) arrived the sides of the stove were red hot, which was much appreciated. All had walked at least a quarter of a mile over rough mountain terrain (some had walked more than a mile) and were quite cold.

SMOKE

"Double, double, toil and trouble; Fire burns, and cauldrons bubble", Shakespeare, 1606.

March 1932

What a sight for a traveler passing Mt. Hall School! On a mountain side, across the way, a slight figure in blue denim overalls, black aviator's leather cap and black high top shoes, was pacing, dancing, hopping, frolicking, jumping, and dodging to avoid smoke, steam, sparks and hot coals that spewed and erupted from a black iron kettle and the wood underneath. To the superstitious folk, I must have looked like a witch's elf tending a cauldron.

Actually I was only making maple syrup. The spring thaw caused the sap to run in the maple sugar trees and by "tapping" we had collected more than a barrel (30 gallons) of it. I had begged for this job. I thought it would be easy and fun – not so back breaking as sawing wood, the other major endeavor of the day.

The acrid smoke caused tears to flow and the wind shifted constantly keeping me forever on the move. Unknown to anyone at the time, the smoke particles contained more than partly oxidized wood. They also contained tens of thousands of chlorinated organic compounds at more than part per quintillion levels and included one called "dioxin". We folk of the mountains were exposed a lot because we had a kettle going two or three days every week. Apple butter boiling, clothes washing, and soap making were done this way. Work on a self-sufficient organic farm was never easy!

MATRICULATION

"This is the thing that I was born to do!"
Samuel Daniel, 1599.

September 1939

"I think I'll major in French", I told Dr. Wright, Dean of Bridgewater College. I was registering a week late and had missed orientation. The Dean, a typical academic, was all business when he asked, "What chemistry course will you take?" I responded, "What is there?" He said, "One designed so that you will never take any more chemistry, and one designed to prepare you to take more." I answered, "Put me down for the first". "I'm putting you down for the second", he said with great authority. So, I took the tough general chemistry course, but refused to synthesize chlorine. The following spring my chemistry professor, Dr. Frederick Kirchner urged me to major in the subject and I concurred. The "A" earned in this first course was most encouraging. Three years later I knew that I wanted to be an industrial analytical chemist.

I would spend most of my college career in Memorial Hall. Chemistry laboratory in the basement; science, mathematics and physics lectures with physics lab, on the first floor; chapel (daily attendance required) on the second floor.

Memorial Hall, Bridgewater College. Lithography by Duane
B. Hylton.

MAGIC

"The chymists are a strange class of mortals impelled by an almost insane impulse to seek their pleasure among smoke and vapor, soot and flame, poisons and poverty, yet among all these evils, I seem to live so sweetly, that may I die if I would change places with a Persian King", Johann Joachim Becker, 1669.

Spring 1942

Reddish-brown fumes poured out of the open windows and doors of Memorial Hall. I anxiously watched from the terraced bank overlooking the athletic field. About a dozen other chemistry students joined me. Many of us had textbooks and continued to study. Our only concern was whether we could re-enter the basement laboratory in time to prevent the aqueous iron solutions we were oxidizing from going to dryness.

We had stayed in the laboratory, where experiments were carried out without the benefit of fume hoods, as long as our respiratory systems would permit. Gasping for air, we had escaped one by one. The entire laboratory soon filled with bromine vapor, which was being used to oxidize iron from the ferrous to the ferric oxidation state.

In this laboratory, we worked unprotected except for ankle length rubber aprons and thus experienced the feel, smell, taste, sight, and noise of chemicals. It may be said that we experienced holistic chemistry. By now I had experienced by my own hand: the biting dehydrating sting of concentrated sulfuric acid, the yellowing of skin from concentrated nitric acid, the swollen degradation of skin from sodium hydroxide solutions, the smothering effect of breathing greenish-yellow chlorine gas, the explosion of hydrogen, the brilliance of a glowing splint in an atmosphere of oxygen, the pungent odor of hydrogen sulfide, the suffocating effect of sulfur dioxide, and the fizz of concentrated acid on organic matter. In such an atmosphere it was easy to appreciate and understand the behavior of molecules.

There was not a doubt in my mind. What happened in a college chemistry laboratory was weird! There was no relationship with the real world. Chemistry was magic – a reaction was the unleashing of forces that were more likely evil than righteous. Of course none of us knew anything about the long-term effects on acute exposure to various chemicals. Research had not yet been done and human experience spoke only to immediate effects. All laboratories known to me, both academic and industrial, had minimal safety precautions, if any. Not until major chemical companies, such as du Pont and Dow developed toxicological research programs did the possible and probable hazards become identified and safety measures instituted.

The real world, I thought, consisted of the semi-superstitious society in which I grew up. Persons living in the Shenandoah Mountain region believed very deeply in God. But they also, at least partly, believed in devils, demons, witches, witch doctors, spooks, and ghosts. Moreover, they believed in doing many things, including planting crops, according to the phases of the moon. Certain decisions were often made based on the observed behavior of black cats, bats, owls, caterpillars, locusts, and various other creatures. The ever dramatic, apparent changing of the environment due to various weather conditions was amplified by the terrain of the mountains, hollows, and creeks adding considerably to the uncertainty of life. Thus, a superstitious outlook was not unreasonable.

My parents and childhood neighbors knew a great deal about the flora of the region. Each species of herb, bush, tree, and weed had a name. Most of these had a specific use. Many different teas were brewed – pennyroyal, peppermint, pipsissewa, teaberry, sassafras, catnip, boneset, and mullen. Each of these provided special relief for various ailments.

These people were equally gifted in the husbandry of animals. My grandfather was typical. On his 200-acre farm, he raised horses, cows, sheep, hogs, geese, ducks, turkeys, chickens, and guinea fowl. To feed these animals and the family, crops of corn, wheat, rye, oats, buckwheat, potatoes, turnips, rutabaga, tomatoes, cabbage, beans, sweet corn, popcorn, cucumbers, pumpkins, and various other vegetables were grown. Fruit trees consisted of many varieties of apples, cherries, plums, damsons, pears, apricots, peaches, grapes, and gages. Growing wild were blackberry, black haw, raspberry, dew berry, red and yellow strawberries, huckleberry, blueberry, gooseberry, elderberry, black walnut, butternut, hickory nut, hazel nut, chestnut, chinquapin, and persimmons.

To accomplish growing and harvesting all of the above required the careful practice of what has later been called, "organic farming". This then was the real world, I thought. A chemist was a sort of witch dealing with potions. No wonder our English professor called us, "silly freshmen".

DECADE III
ADJUSTING

THE ROANOKE RIVER

*"Man comes and goes but while he's here he fills the waste baskets
higher and higher.
It ain't right", Hans Opperheimer, 1964.*

Summer 1942

The small wiry bearded man strode confidently down the hill, climbed the rail fence, and proceeded up the road and onto the bridge where I was at work taking samples of river water. "Howdy!" he greeted me. "Nice un, huh? It's time ya' people did sumpun." (He thought I was an employee of the state. It's just as well, I thought. No telling what he might do if he knew I was a chemist with American Viscose Corporation.)

"What's going on in Russia?" he asked. "I'm sorry, I don't know." I answered. "Well, if you wuz a child o' God, you'd know. I'm a child o' God and I say tuh ya' the Germans are outside Stalingrad. The Bible prophesy is being fulfilled. The Russians are th' red beasts and the Germans are th' black. I say tuh ya', its Revelation, all is tuh it. 'Cause Christ rode in Jerusalem on th' jawbone o' his ass." He went on and on and only quit when I, quite by chance, said "Amen".

It was a scene to be repeated every working day of the summer. The sermons varied only slightly in content. Constantly I was admonished to repent and be baptized. Constantly, I was encouraged to find the "rot in the river" and force the company to quit "pouring crap" into it. It was my first encounter with an aspiring environmentalist!

A study of the Roanoke River was one of three projects assigned to me during summer employment at the Roanoke plant of the American Viscose Corporation. In the morning I sampled the river from a bridge upstream from the plant and in the afternoon, from a bridge about five miles downstream from the plant. Part of the analytical procedures were carried out on the bridge and part back in the laboratory on my return. The chemical reagents were carried in a small black bag not unlike those used by physicians making house calls. Samples were taken by lowering a lead bucket, onto which two bottles were fastened, into the river to the depth of interest. By using bottles of different sizes, water could be siphoned into them.

I had a partner named Bob who was one of the "Five Smart Boys" from Roanoke College. The "Five Smart Boys" were a basketball team that became national heroes after a superb performance in the NIT tournament in Madison Square Garden. Bob viewed himself as my supervisor and went on sampling expeditions with me, but only in good weather. Each time he went he explored the riverbank only to startle me with screams of terror whenever he encountered a plant, which he thought was poison ivy; or a snake, which he thought might be a poisonous water moccasin. Bob was one of many people I have known who was terrified by anything described as poisonous or labeled with a skull and crossbones. So, his behavior was not uncommon, but was typical of many that had grown up in town and had little opportunity to commune with nature. Rufus was the uniformed chauffeur who drove me to the sampling bridge in a sleek black Buick limousine, he often had to rescue Bob from his exploration ordeals.

We were primarily interested in the oxygen content and pH of the water. Water upstream of the plant usually had an oxygen content of about six parts per million, but only about two downstream. This difference was due to the waste sulfite solution, which was in the effluent from the plant. pH is a way to measure very small amounts of hydrogen ions and hydroxyl ions. pH can range from 1-14. pH of seven means that the concentrations of these two ions are equal. Less than seven is acid and greater than seven is alkaline or basic. Upstream of the plant the pH varied from about six to nine. Downstream the pH was from three to six. Both oxygen content and pH varied significantly with weather conditions, both being higher after a rainstorm.

We also measured oxygen consumed and biochemical oxygen

demand. These gave us a measure of the amount of waste viscose polymer escaping to the river. We could, however, see that it was considerable as the rocks in the river were covered with a brown moss-like material, which waved in the river's current. The appearance of the river was only to have been expected in a horror movie. Except for a few minnows in the shallows, no life was observed in this part of the river. This was in great contrast to the river upstream of the plant where large carp were always swimming under the bridge and the only sign of pollution were the numerous condoms floating on the water.

As we sampled the river, cars stopped and people watched the proceedings carefully. Traffic was often blocked until I had closed my case and started walking away. Everyone assumed I was an employee of the state and asked very few questions, probably fearing to appear stupid. Apparently, there were no reporters among them. If there were, newspapers refused to publish their stories.

It was apparent to me that although the citizens preferred a clean river, they were willing to sacrifice it for the jobs that the plant provided.

By the end of the summer, I had made two major decisions. First, I wanted to become an industrial analytical chemist. Second, I was more aware than ever of our environment and wanted to make a contribution to its improvement. Both, however, would need to be done on a firm scientific basis, I thought. I had reason to think so because my alarmist letters to newspaper editors went unpublished. Science and the analytical laboratory had become my real world.

CHLORINE ALONG THE JAMES

"We are confronted with insurmountable opportunities",
Bill Gore.

July 1943

I felt as though I was being led by a spelunker through an enormous dimly lit cavern, with walls oozing watery substances, which dripped on my face and shoulders, and formed puddles on the floor. I had no way of knowing which drop was acid and which was not, adding to my already queasy weird feeling. What an utterly horrible place, I thought. But it was not a cavern; it was the ground level of the nitric acid production plant of The Solvay Process Company in Hopewell, Virginia. The assistant director of the Quality Control Laboratory was taking me to my first job assignment in the Chlorine Production Plant Control Laboratory.

The chlorine plant, below City Point on the James River, was even more of a shock. Twelve stories high, built of steel girders, all highly corroded, some almost rusted through, steam and suspected noxious gases escaping from most tanks and many pipe lines, the plant looked like a high tech evil cauldron producing a mishmash of variously colored fumes. Yet, in the middle of all this decay was a silvery shining pipe of tantalum untouched by the oxidative acidic environment. It was this pipe that contained the chlorine and nitrosyl chloride.

I was scared. It appeared obvious to me that this was a high-risk place to work. Chlorine and nitrosyl chloride leaks in the plant occurred frequently. One day I unwittingly stepped into a pocket of such gas. The choking impact knocked me to the floor and I had to report to the health department for an oxygen inhalation treatment.

Almost all procedures on this job were hazardous. The most obvious were the procedures for sampling the chlorine distillation column, the chlorine lines for loading, and chlorine cylinders. The same technique was used for all these situations. Using a portable gas mask and neoprene gloves, I allowed liquid chlorine to purge through a glass trap immersed in a Dewar flask containing dry ice in acetone. Sometimes the chlorine would spill through the trap evaporating in a yellow cloud. One could only pray that the canister of the gas mask had enough capacity to absorb all the chlorine coming one's way.

An unexpected hazard was discovered when returned "empty" chlorine cylinders were sampled. Usually these cylinders still contained a few pounds of chlorine. By analyzing this chlorine we could determine if it had been produced by Solvay or by some other company. Sometimes the valves were frozen shut and could possibly be broken in the attempt to open them or sometimes they opened easily but then could not be closed. In the latter case the only recourse was to leave the vicinity as quickly as possible and allow the gas to vent to the atmosphere. These were enough hazards, but to add to the excitement, black widow spiders were often found lurking under the metal covers protecting the valves.

Chlorine was shipped by barge from this plant to Wilmington, North Carolina, where it was used by The Dow Chemical Company to extract bromine from seawater. Great care was taken to avoid having chlorine leaks while the barge was in route. So the cylinders were tested for leaks at the plant loading dock and loaded on the barge. The chlorine control chemist had the job of inspecting cylinders on the barge. Generally no leaks were found, and those found could almost always be stopped by tightening one or more valves. However, one day I detected a leak larger than usual. The foreman of the loading station tightened the value but the leak persisted. The decision was made to return the cylinder to the plant to unload it. A railroad crane with a flat bed was employed and the cylinder was placed on the flat bed. The foremen sat astraddle the cylinder to prevent it from rolling off the bed and the crane moved quickly to the plant. On arrival the normally concave ends of the cylinder were now convex, and the cylinder was found to contain two tons of chlorine – twice the amount for which it was designed. Such casual handling of situations, which were potentially lethal, was typical of the time.

How could any college graduate allow himself to be placed in such a position? After the bombing of Pearl Harbor, the United States mobilized for war. Many young college men, not trusting the Selective Service System, volunteered for officer training. Those students in good standing with scientific and technical training, however, were often deferred by the local draft boards so that they could complete their college training before entering the military where they could undertake special assignments. Roger Barnhart and I were such students. After graduation we had about two months left on our deferments, although it had been announced that all able bodied men between the ages of 21 and 28 would be drafted. We were both able bodied and would have to go. So we sought commissions in the military but all the training programs had been filled. Rather than remain idle we applied and were awarded jobs in one of the most war-related industries in Virginia, The Solvay Process Company. Once there we were "frozen-on-the-job" and had to obtain a release from the company before we could be considered for a commission by the military.

From the beginning I was considered a pair of hands by the management, but portrayed as a "one-of-a-kind" scientist to the Selective Service System. I was initially assigned to the Central Analytical Laboratory where the conditions were deplorable. With June temperatures approaching 100°F and no air conditioning, we stripped to the waist and performed the same task over and over each day. I worked with an older chemist. One day I would weigh samples all day and he would titrate them. The following day the roles would be reversed. Initially I could not keep up with his fast pace, but since I was younger and had more endurance, soon he could not keep up with me. Apparently this was sufficient for the laboratory director. He came by and informed me that I was no longer "on probation" and would be "rewarded" with my own control laboratory in the chlorine plant working 6 days a week with a shift schedule of the first week working 8 a.m. – 4 p.m., the second week 4 p.m. – 12 midnight., and finally the "graveyard" shift, from 12 midnight to 8 a.m.

Gradually I became accustomed to the chlorine plant and no longer saw the hideous dangers. I could take samples of liquid chlorine from any line or cylinder without being anxious for my life. I could walk out on the sheet metal chlorine-sampling platform, suspended in space from the eleventh floor and supported only by two highly corroded steel braces, without hesitation or pausing for prayer. I could repair leaks in the gas analysis system in the laboratory while breathing a little chlorine gas without panic. In short, familiarity made the place seem friendly. Fortunately, my safety habits had been made firm while I was still fearful and distrustful of the place.

The air in the chlorine plant not only contained traces of chlorine, nitrosyl chloride, hydrogen chloride, and nitrous oxide from the plant itself, but ammonia from the ammonia plant across the road and nitrogen dioxide and nitric acid from the nitric acid plant next door. Strangely, I don't recall that fly ash was a problem in this plant, although, the company had its own powerhouse. The noxious fumes often dulled ones senses but a release of ammonia would waken us up again.

Although I could usually finish all the chemical analyses for plant control purposes within three hours of an eight-hour shift, I was not allowed to analyze the air or to develop methods for doing so. All analytical procedures came from the research laboratory at headquarters in Syracuse, New York. None of these were designed to determine occupational exposure of personnel to chemicals or environmental contamination of the community of Hopewell.

The chlorine plant was not a place for visitors. This was fine with me because I would not have had enough courage to expose relatives and friends to the threatening manifestations of the plant.

Something needed to be done about the workplace and the environment. No one working for Solvay could doubt that. A National Institute for Occupational Safety and Health and a Federal Environmental Protection Agency were possible future programs to dream about.

Because of the hostile environment, I was determined to resume my education as soon as I could be released from the job on which I was frozen by the federal government. However, the course I should take was not clear. Then the company started to do research on organic detergents and a young chemical engineer named Olson from the University of Minnesota, in charge of the work, became a friend of mine. Soon I learned that no chemist could determine the purity of such products nor measure any organic contaminants in the product. I resolved to learn to determine organic compounds in other organic compounds. (I was thinking of percentage levels.)

A few months later, as a new graduate student at the Ohio State University, I learned that very little was taught about the analysis of organic compounds at that or any other university. To cover every possibility, I took courses in thermodynamics, kinetics, quantum chemistry, organic mechanisms, qualitative organic analysis, organic stereochemistry, aliphatic organic chemistry, aromatic organic chemistry, electrochemistry, spectroscopy, instrumental analysis, colloid chemistry, and oxidation-reduction potentials. In addition, over a period of five years, I helped teach almost all the courses in analytical chemistry offered by the university. Even so, I knew how to measure only a very few organic molecules at levels less than 1 percent (ten billion parts per trillion) when I joined The Dow Chemical Company in 1951.

Today, I am told many non-Dow chemical plants in third world countries are similar to this one in which I worked.

THE UNINTENDED PLEDGE

"He who enters a university walks on hallowed ground",
James Bryant Conant, 1936.

Spring 1947

"No one can pigeon hole me – not even you!" I stormed at Professor Edward Mack, Chairman of the Department of Chemistry at Ohio State University. I was in his office to protest the letter I had received informing me that I had permission to earn a M.S. degree and then they would decide if I could try for the Ph.D. "I do not accept this insult! I am going directly for the Ph.D."

I had just surprised myself! I had decided, while in industry, that there would be great opportunity for organic analytical chemists in industry and that the master's degree was best suited to take full advantage of this. I was not interested in a chemistry doctoral program. But, I was in for an even bigger surprise. "That's the best news I've had all week!" Dr. Mack replied with a tear-filled smile so wide that it caused ashes to fall from the cigarette held in the middle of his lips. Thus, I was bound by my word and now had to work even harder to prove myself.

Four years later as I received the Ph.D. degree, Dr. Mack urged me to accept work at the Oak Ridge National Laboratories where he was a member of the Board. But I was fearful of working with toxic substances such as radioactive ions and refused the offer.

SOOT AND SNOT

"Youk'n hide de fier, but w'at you givine do wid de smoke?,
Joel Chandler Harris (1848-1908), Uncle Remus, Plantation
Proverbs.

Winter, 1949

"What in the world happened to you?" I was greeted by my fellow graduate student and lab partner. "Nothing", I was pleased to reply. "You'd better look in a mirror", he said. "You look like you've been in an explosion."

So I looked in the mirror! Wow! There were black smudges on my face, hair, white shirt and jacket. I tried to brush them off with tissue papers, but they only spread and became smears.

Then I felt the need to sneeze and blow my nose in a large clean white handkerchief such as all men carried at the time. The result was a ruined handkerchief – soot suspended in mucus blackened the cloth. I cleared my throat and spit out more soot. Coughing, sneezing, blowing, spitting, I finally succeeded in ridding myself of soot. It was an astonishing experience.

However, I should not have been surprised. With my thoughts on the experiments of the day, I had strolled under many trees on the three-block walk to the laboratory. The branches and twigs on the trees were covered by a layer of soot, and when a breeze came, the tree branches and twigs dropped clusters of soot particles to the ground. Some had apparently found me.

The soot came from the chimneys of homes that burned soft coal. About 300,000 people in the city were warmed by the burning of soft coal.

Unknown to us at the time, these soot particles probably contained at least a part per quadrillion of "dioxin" – enough to make modern environmentalists cringe in fear and propose dire measures to ban soft coal.

DECADE IV
TUNING UP

INTERVIEW PRIORITIES

"Man always travels along precipices. His truest obligation is to
keep his balance",
Pope John Paul II

October, 1950

We stood on the banks of Dow's waste water treatment facilities, Mr. Alonzo W. Beshgetoor and I. I pondered my future and wondered what this series of lagoons and aeration ponds had to do with the practice of analytical chemistry. Then Mr. Beshgetoor, a man of few words, spoke, "It is one of the wonders of the world", he said. "Yes, it has to be!" I replied, after he had explained that waste chemicals were eaten by bugs (bacteria) especially trained to do the job. I inhaled deeply and detected a faint, pleasant odor of phenolic compounds (some probably chlorinated) together with a little suggestion of bromides. Chlorine was not detected and other odors were not noted.

I had met Mr. Beshgetoor just 20 minutes before at the Technical Employment Office where he had appeared with tan raincoat and fisherman's hat dripping like a lovable character in a favorite fairy tale. "Where's Dr. Stenger?" I had inquired. "Vernon Stenger is the greatest industrial analytical chemist in the world", he said simply. "If he is the greatest, then why are you meeting me?" "I'm the Director of the Main Laboratory", he replied almost as if he were in awe of the title.

We broke our interlude, got back into his 1947 Plymouth Coupe and drove to the Spectroscopy Laboratory where I spent the rest of the morning and much of the afternoon. It was evident that Dow's spectroscopists were very advanced in their field. Late in the after-

noon we took a fast walk through the Main Laboratory with its mountains of glassware coated with ammonium chloride. These laboratories reek with opportunity for improvement, I thought.

That night, I told my wife that I saw unlimited opportunity in
the Main Laboratory and that Dow appeared more concerned with
human health and the environment than I had been able to detect in
other interviews. But I also said that I saw no chance for a job offer
because I had been too frank with Mr. Beshgetoor. I had let him
know that I found the Main Laboratory to be behind in its use of
instrumental methods and we would have to develop or purchase
much analytical equipment if the laboratory were to advance to the
state of the art. He had replied that, "I have gotten everything I ever
asked for." That response puzzled me.

"So relax!" I told my wife. "We will not be moving to Midland."

When the offer came, it was to work with Dr. Stenger in a new
air-conditioned laboratory developing new analytical methods. I decided to try it for a year, considering it a post-doctoral experience.

The problems that year were so interesting and challenging that
I became committed for life!

Dr. Vernon A. Stenger tells how he trained me in his new
"Special Services" Laboratory.

BUTTON AND CREAM

*"Something hidden. Go and find it", Kipling (The Explorer),
1903.*

1953-1954

Among the many analytical problems brought to my attention
by Dow scientists, was the need to determine "trace" amounts of
impurities in products, especially organic compounds. For example,
the purity of organic compounds could be measured by freezing point
curve and Dr. Daniel Stull needed to know the purity of compounds
in order to properly calibrate his state-of-the-art instruments. At that
time we could, in favorable cases, measure concentrations in parts per
thousand. While this example was of great potential utility, a problem arose that was urgent and required an immediate solution.

The latex plant produced a product which had excellent properties for use in paint. But each spring, usually in May, two unwanted
bi-products were found. "Button" consisted of large agglomerates of
latex particles, which could weigh as much as a ton, and were thus
too bulky to pass through the 1-feet diameter orifice at the bottom of
the reactor kettles. "Cream" consisted of low-density latex particles of
latex, which became apparent after paint was formulated. As with
cow's milk, the "cream" rose to the top. No matter how the paint was
stirred, the "cream" caused brush marks when paint was applied to a
surface.

The problem was so acute that Dr. Ray Boundy, Director of
Research for Dow, decided to release the energy and knowledge of
the entire research organization to find a solution. To accomplish this
a research team consisting of a research person from each of seven
different laboratories was organized. I was named chairman of this

group. The work of the group had top priority to have our research programs carried out and we were allowed access to all information and testimony, which would expedite the investigation.

Previously, studies on the latex reaction using pressure bottles demonstrated that trace (part-per-million) levels of many different types of impurities in the reactants could produce effects similar to those called "button" and "cream".

After reviewing this information we began a comprehensive search for trace levels of aldehydes, ketones, oxides, peroxides, oligomers, heavy metals, etc., in the monomers; metals, sugars, organic acids, suspended particulate matter, diatoms, etc., in water; and heavy metals in the detergents. Our search required a review and investigation of all analytical methods to determine these components and visits to other Dow production sites to compare their experiences. From this total effort we were able to demonstrate that the problem came from a synergistic effect of a metal in one of the monomers together with sugars in the water. The concentration of the heavy metal remained constant during the entire year at a level insufficient to produce "button" or "cream". However, when the ice on Lake Huron melted in May, the lake "turned over" (cold water from the melted ice sank and the warmer bottom water rose to the top). The concentration of sugars increased in the water intake. The increase in sugar concentration was sufficient to enhance the behavior of the heavy metals already present.

We eliminated the heavy metal from the monomer and the "button" and "cream" phenomena disappeared. This allowed Dow to build latex plants around the world knowing that they were really not subject to witchcraft as some had previously jokingly (I think) suggested.

This was the kind of problem-solving I liked to do and I hoped to make it an entire career. It was not to be!

SAFETY FIRST

"Science is a willingness to accept facts even when they are opposed to wishes",
B. F. Skinner.

Spring 1958

"So it's impossible that 1,2,4,5-tetrachlorobenzene is leaking into the work atmosphere" the Research Director, Dr. Noland Poffenberger, concluded. He was finishing a talk describing the engineering features which had gone into the design of Dow's chlorobenzene plant.

All eyes turned to me, for I had reported finding this material, highly toxic to animals and humans, in a sample of air taken at that plant by an industrial hygienist. I was intimidated because I was a relatively new employee and this statement was made before a major production manager, Bill Williams, in the presence of several other Dow scientists. Nevertheless, I defended my work by explaining how an ultraviolet spectrophotometer works, showing charts of the raw data and explaining its meaning. Immediately, Dr. Poffenberger changed his mind, declared the evidence as being irrefutable, and advised Mr. Williams to authorize funds to eliminate leaks in the plant. Mr. Williams concurred. Although the cost was very high, the leaks were stopped and Dow workers were not exposed to this harmful substance.

This must have been a very painful decision for Manager Williams. The company was so frugal that all employees, including the president, punched a time card and I had to account for every 20 minutes of my time. But, this was very reassuring for me because it provided proof that The Dow Chemical Company put environmental and human health concerns ahead of profits. I was now completely convinced that I had made the right decision. My choice of Dow Chemical as my employer was the right one. Although I was a newcomer to the department making chlorobenzenes, management was ready to accept scientific evidence and act on it.

This attitude has prevailed at Dow through the 37 years I worked for the company. Furthermore, I have documented evidence that Dow used state-of-the-art analytical methodology to measure various pollutants in its effluent to the river since 1929.

Almost always the management has accepted the reports of scientists and has responded accordingly. For example, in the case of 2,4,5-trichlorophenoxy acetic acid (2,4,5-T), Dow scientists were asked repeatedly if they saw any scientific reason to cancel the manufacture of this material. Each time, the scientists were in unanimous agreement. There was no scientific evidence, which mandated discontinuance of the product. However, defense of the product became too expensive and the Dow manufacturing plant was shut down. Later Dow withdrew from negotiations with EPA. Most scientists were irate. After all, they had been defending a principle, and the data reported in thousands of scientific articles support that principle.

THE DEBUT

_"From ghoulies and ghosties, and long leggity beasties, and things
that go bump in the night,
Good Lord, deliver us", Old Scottish Prayer._

Fall 1964

In my office, V. K. Rowe, Dow's chief toxicologist, advised me that tests conducted in his laboratory had shown that a molecule, 2,3,7,8-tetrachlorodibenzene p-dioxin, was a powerful toxin in guinea pigs. This molecule could possibly be present in wastes from various plant processes in which the products were chlorinated aromatic hydrocarbons. In order to protect the workers, we needed to know which wastes contained "dioxin" and how much was there. I agreed that my analytical group would fractionate waste oils and have him test each fraction for toxicity. Dr. Norman Skelly fractionated the tars and identified the components. The one with the highest concentrations of 2,3,7,8-TCDD was the most potent acnegen. In this manner, dioxin first appeared to me!

"Dioxin" makes its presence known by appearing at accidental gatherings of its relatives around the world. It was present, but remained anonymous, at an industrial accident in a Monsanto trichlorophenol plant in Nitro, West Virginia.[1] This accident resulted in 122 cases of chloracne* among the workers. If it were known that dioxin caused the skin problem, such knowledge was not reported.

In 1953, in Ludingshafen, Germany, dioxin again "crashed" a party. This time, 55 BASF Aktiengesclschaft workers developed chloracne.[2] A search was made for the agent causing chloracne. 2,3,7,8-Tetrachlorodibenzo-p-dioxin, together with tri- and tetrachlorodibenzofuran, was found by investigators at the University of Hamburg-Eppendor[3]. Dioxin had made its debut.

In 1963 in Amsterdam, the Netherlands, an explosion in a NV Phillips herbicide factory[4] exposed 145 people, 69 of whom developed chloracne.

In 1964, sixty-one Dow workers were unintentionally exposed to waste oils and tar in a trichlorophenol plant during a process change. Of those, 49 developed evidence of chloracne. On Dr. Rowe's advice we decided to identify the active agent.[5] Dr. Norman Skelly succeeded in isolating a material which produced chloracne on rabbit ears with great sensitivity. This material was then identified by a series of sophisticated analytical techniques as being 2,3,7,8-tetrachlorodibenzo-p-dioxin. This material quickly became known in the media by the simple term "dioxin". We proceeded to develop a method to measure the "dioxin" content of trichlorophenol and related materials. Harold Gill and co-workers did this by using gas chromatography with flame ionization detection. Unknown to us, David Firestone[6] at the United States Food and Drug Administration had simultaneously developed a method to determine other chlorinated dioxins in fat from chicken feed. His method was very similar to Gill's, but he used electron-capture detection. His work was necessary because higher chlorinated dioxins, namely some members of the hexa—and hepta—congener groups had gotten into chicken feed and caused excessive fluid in the heart sac and abdominal cavity in chicks fed toxic fat. The resulting condition became known as chick edema disease and the activity of the feed was called the "chick edema factor".

An explosion in a Coalite and Chemical Products, Ltd., plant at Bolsover, United Kingdom in 1968 resulted in 79 workers developing chloracne.[7]

Waste oils were used for dust control on southern Missouri roads and horse arenas in 1971. In the horse arenas, a number of animals died. In one of the arenas, about 33 parts per million TCDD was found in the soil. In another arena, two children developed a disease resembling chloracne.[8] The mismanagement of waste oil in this way was a terrible mistake, but the evil was made much worse by the use of such soil as topsoil in Times Beach and Imperial, Missouri[9]. The resulting public outcry was so severe that the U. S. Environmental Protection Agency bought the entire town of Times Beach.

As a result of an industrial explosion, about seven square miles of countryside around Seveso, Italy, was contaminated by chemicals, including TCDD, on July 10, 1976. The resulting panic was felt by the entire world.[10, 11] As many as 30,000 people were exposed. One headline read, "Toxic Cloud Chases 1,000 Italians". The heaviest contamination involved an area of 180 acres, occupied by 736 people. Many of these people were evacuated from their homes. Some women had therapeutic abortions, but the fetuses were normal. The most contaminated earth was bulldozed into a mound and encased in concrete.

In the Seveso case, as in all the other accidents, the only lasting consistent finding of physical harm was chloracne.[12] Chloracne appears as a severe case of teen-age acne. It covers the face, ears, and back of the victim. Early reporters described it as a horrible grotesque disfiguring affliction which could result in death. Cases I've seen, however, appear to be no worse than a bad case of acne vulgaris. By 1954, scientists reported that the conditions could be caused by chlorinated biphenyls [13] and by 1987 they knew that 2,3,7,8-tetrachlorodibenzene-p-dioxin and 2,3,7,8-tetrachlorodibenzo-furan were much more potent.[14] By 1983, mortality data indicated that chloracne had no effect on life expectancy.[15] Nevertheless, dioxin had dramatically introduced itself to human society.

References

1. J. A. Zack and R. S. Suskind, "The Mortality Experience of Workers Exposed to Tetrachlorodibenzodioxin in a Trichlorophenol Plant", J. Occup. Med, 1, *11*, 1980.

2. P. J. Goldman, "Severe Acute Chloracne: A Mass Intoxication by 2,3,7,8-tetrachlorodibenzodioxin", Der. Hautarzt, *24*, (4), 149, 1973.

3. J. Kimmig and K. H. Schulz, "Berufliche Akne (sog. Chlorakne) durch chloriete armatische zyklische Äther", Dermatologica, *115*, 540, 1957.

4.L. M. Dalderup and D. Zellenrath, "Dioxin Exposure: 20-Year Follow-up", Lancet, 1134, 1983.

5.W. B. Crummett and R. H. Stehl, "Determination of Chlorinated Dibenzo-p-dioxins and Dibenzofurans in Various Materials", Environ. Health Persp., 5, 15, 1973.

6.D. Firestome, "Etiology of Chick Edema Disease", Environ. Health Persp., 5, 59, 1973.

7.G. May., "Chloracne From the Accidental Production of Tetrachlorodibenzodioxin", Br. J. Ind. Med., 30, 276, 1973.

8.C. D. Carter, R. D. Kirsbrough, J. A. Little, R. E. Cline, M. M. Zack and W. F. Barthel, "Tetrachlorodibenzodioxin: An Accidental Poisoning Episode in Horse Arenas", Science, 138, 738, 1975.

9.Missouri Dioxin Task Force, Interim Report, 1983.

10.E. G. Homberger, J. Sambeth, and H. K. Wipf, "The Seveso Accident: Its Nature, Extent, and Consequences", Am. Occup. Hyg., 22, 327, 1979.

11.F. Fanelli, et al., "TCDD Contamination in the Seveso Incident", Drug Metabolism Review, 13 (3), 407, 1982.

12.M. Donelli, "Seveso, inflation of fear", Chemtech, 19, 139, March 1989.

13.J. W. Meigs, J. J. Albom, and B. L. Kartin, J. Am. Med. Assoc., 154, 1471 (1954).

14.J. Kimmig and K. H. Schulz, Dermalotogica, 115, 540, (1957).

15.L. M. Dalderup and D. Zellenrath, the Lancet, Nov. 12, 1983, pp 1134-35.

CHRISTMAS INTERRUPTUS

"Heap on more wood! The wind is a chill;
But let it whistle as it will,
We'll keep our Christmas merry still",
Marimon (1803) VI, Introduction, St. I.

27 December 1969

Christmas at our house is celebrated in the main stream American tradition, generally consuming all the time from December 20 through 31. Symbolism from many nations is blended into a gigantic environment of love and peace. It is a time for family and friends, a time for joy, but also of introspection. Christmas 1969 promised to be no different until, on December 20, I was told that Dow was asked to report to the Office of Science and Technology what we knew about the manufacture, analysis, toxicology, and epidemiology of the herbicide 2,4,5-trichlorophenoxy acetic acid (2,4,5-T) and its esters. The meeting was scheduled for December 27 in Washington, DC. This meant that preparation would have to be made over Christmas itself.

My responsibility was to present the analytical information on the product. We had analyzed the acid for total composition by three distinctly different methods: infrared spectrophotometry, ion exchange chromatography, and gas chromatography. Data taken by the three methods was in substantial agreement and I was proud to present and discuss it.

Members of the Office of Science and Technology appeared restless. They kept interrupting me to ask when I was going to talk about the impurity known as "dioxin". Relentlessly, I gave them a complete material balance on the composition of 2,4,5-T acid. When I finally explained the new methodology which Dr. Rudolph Stehl of our laboratory had just developed to determine 2,3,7,8-TCDD in technical products, they were much more attentive. Rudy's finding of 27±8 ppm 2,3,7,8-TCDD in the Bionetics sample was readily accepted. Perhaps they were charitable because of the holiday season, or perhaps they were favorably impressed by the elegance of the technology. More likely they were relieved that they could make a good

Dr. Rudolph Stehl found and measured 2378-TCDD in a competitor's product.

report to the President, Richard M. Nixon. In any case, I was not subjected to the usual academic crossfire that professors like to ignite around research reports. I explained to them that this measurement had been made by a new method which had not been fully validated, and the number had not been confirmed. Nevertheless, the number was later used by toxicologists in teratology studies (Courtney, K. D. and J. A. Moore, Toxicol. Appl. Pharmacol., 20, 396, 1971).

The Bionetics Laboratory had studied the teratogenic effects of various pesticides on mice. Among them was the herbicide 2,4,5,-T. The sample had been "pulled off the shelf", as was common practice among toxicologists at that time (Its composition was not considered important.). Later, it was identified as a competitor's technical material, and we found it to contain dioxin at a level considerably higher than normal. If Bionetics had selected a Dow material or that of almost any other manufacturer, the events recorded in this book would probably not have happened as the teratogenic effect would not have been detected. From this experience, toxicologists learned that material of known purity should be used in such studies.

So, "dioxin" interfered with Christmas. Working over Christmas was no big deal, but the associated stress of the first time appearance before so distinguished a body of scientists prevented me from experiencing the true spirit of Christmas. While I stood before the group described as being "only one heartbeat from the president", I felt that a molecule, "dioxin", was affecting me in a supernatural manner.

DEFINING DIOXIN

"I hate definitions", Benjamin Disraeli, 1804-1881
*"To define is to exclude and negate", Jose Ortega y Gasset, 1883-
1955*
*"It is not the facts which divide men but the interpretation of the
facts", Aristotle.*

The scientists with whom I was closely associated, locally, nationally, and internationally, strove mightily to produce data which were irrefutable and unbiased. Almost always they were able to achieve this goal. The analytical data met or exceeded the "guidelines" set by the American Chemical Society and the Environmental Protection Agency. The toxicological and epidemiological data set new standards for excellence and integrity and was reviewed by many scientific authorities. All these measurement scientists agreed the data were sound and free of bias. Never before had the behavior of a molecule been studied so thoroughly at all concentration levels with consensus of scientists of almost all industrialized nations.

This extensive body of data was now available to decision makers for their final word. Regulations came fast and furious. In the United States they were almost always set at absurdly low levels, reflecting the fear of the community, state, or the nation.

Once suggested, the regulations could not be relaxed no matter what any new data showed. But, sooner or later, political leaders of all governments set up a "study" group to assess any concerns anyone may have. Scientists are seldom involved. Concerned citizens, newly given the opportunity to consult with panels, task forces, or committees on matters involving "dioxin" almost always demand an immediate definition of the term.

My review of six well-known dictionaries reveals five different definitions. However, only The Oxford English Dictionary, 1989, has all statements in accord with sound scientific facts, but even it is incomplete. Webster's Third New International Dictionary 2000, does not list "dioxin" and perhaps that is best since the term is so very difficult to define in a few words. Likewise, Encyclopedia Britannica, 15th Ed., Vol. 4, 2002, is the only one of three encyclopedias which presents a discussion free of speculation and bias.

Obviously "dioxin" is extremely difficult to define. Any definition is unavoidably influenced by the background of the author. For example, if I wrote the definition I would emphasize points of interest of analytical scientists. The following might result:

"*Dioxin* is the scientific name of each of two non-toxic isomers of the heterocyclic molecule, $C_4H_4O_2$. Therefore, it is also the base name [surname (?)] for a vast family of molecules built around the $C_4H_4O_2$ structure. The most prominent congeners are the chlorinated dibenzo-p-dioxins because of their toxicity in animals and the Herculean efforts of some governments to eliminate their existence. The most feared of these is 2,3,7,8-tetrachlorodibenzo-p-dioxin, (2378-TCDD), which is commonly referred to as 'dioxin'. First discovered by industrial scientists as an unwanted trace byproduct in an effective defoliant, 'dioxin' was first thought to result only from anthropogenic activity. Recent evidence, however, strongly suggests it has also been produced naturally. For example, it is one of thousands of different molecules produced at ultra trace levels in a fire."

To this one could add any behavior of "dioxin" of interest to the study such as a chemical stability, thermal stability, light stability, bioaccumulation, bioconcentration, adsorption, absorbtion, solubilities, environmental movement, toxicity in organisms, etc., etc. These should be added only if the data is proven scientifically valid. To properly appreciate the significance of such information requires a minimum knowledge of environmental chemistry as taught by Prof. Nigel J. Bunce in "Introduction to Environmental Chemistry", Wuerz Publishing Ltd., Winnepeg, Canada, or equivalent. More importantly, the uncanny ability of "dioxin" to create fear in people should be thoroughly explored. In this book, enough information is presented for the reader to add considerably to the definition.

Unfortunately, the interested citizen is likely to discover one of many government statements designed as evidence to further the fear of "dioxin" such as the one in the May, 1997, issue of *Environmental Health Perspectives* which is more concerned with "links" than scientific facts.

THE FAMILY NAMED DIOXIN

"What's in a name? That which we call a rose by any other name
would smell as sweet.",
Shakespeare, Romeo and Juliet, II, 2.

"He left the name at which the world grew pale", Samuel
Johnson, Ib.1. 221.

The name "dioxin" has become confusing. The popular media calls 2,3,7,8-tetrachlorodibenzo-p-dioxin simply "dioxin". However, when the term "dioxin" is used several, and sometimes all the chlorinated dibenzodioxions are meant. Worse still, some scientists, especially environmental scientists, also adopt this lazy language. Such usage is in complete disregard for international scientific nomenclature, which has been carefully developed and used by scientists in all nations of the world.

Scientifically, there are only two dioxins. They are 1,4—or p-dioxin and 1,2—or o-dioxin. They consist of four carbon (C) atoms and two oxygen (O) atoms in a ring with hydrogen (H) atoms attached to each of the carbon atoms. The structural formulas look like this:

1,4-Dioxin or
p-Dioxin

1,2-Dioxin or
o-Dioxin

Scientists use an abbreviated structure, which is as follows:

1,4-Dioxin or
p-Dioxin

1,2-Dioxin or
o-Dioxin

Scientists often abbreviate 2,3,7,8-tetrachlorodibenzo-p-dioxin to 2378-TCDD. This is the same molecule which the popular press calls "dioxin". However, it is entirely different from the 1,4-dioxin or 1,2-dioxin identified and named scientifically. The unchlorinated basic dioxin compounds consists of three six atom-membered rings fused together. Two rings consist of six carbon atoms each and are benzene (benzo-) rings. One ring consists of four carbon atoms and two oxygen atoms and is similar to the ring of 1,4-dioxin, also known as p-dioxin. The carbon atoms outside the dioxin ring have hydrogen (H) atoms attached to them. The structure thus far developed is shown as follows and is called dibenzo-p-dioxin.

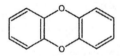

Chlorine (Cl) atoms can replace hydrogen atoms on the ring. From one to eight of the hydrogen atoms can be replaced. Also these can be replaced in a number of different positions. When one chlorine atom is involved, the resulting molecule is called monochloro— and the one chlorine atom can occupy either one of two different positions. The resulting two molecules are called isomers. Other numbers of chlorine substitutions are named and result in different number of isomers as follows:

Number of Chlorine Atoms	Number of Possible Isomers	Name of Isomer Group	Acronym
1	2	Mono	MCDD
2	10	Di	DCDD
3	14	Tri	T$_3$CDD
4	22	Tetra	TCDD
5	14	Penta	PCDD
6	10	Hexa	HCDD
7	2	Hepta	H$_7$CDD
8	1	Octa	OCDD

The chlorine atoms may occupy eight different sites on the molecule. These sites are numbered in order to distinguish iosmers and to name them appropriately. The numbering is as follows:

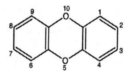

Therefore, 2,3,7,8-TCDD has the following structure:

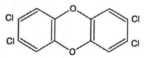

The chlorinated para-dioxin family is as follows:

DIOXIN

p Dioxin or 1,4-Dioxin

DIBENZODIOXIN

Dibenzo-p Dioxin

THE MONOCHLOROS

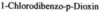

1-Chlorodibenzo-p-Dioxin

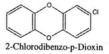

2-Chlorodibenzo-p-Dioxin

THE DICHLOROS

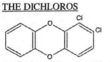

1,2-Dichlorodibenzo-p-Dioxin

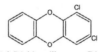

1,3-Dichlorodibenzo-p-Dioxin

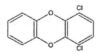

1,4-Dichlorodibenzo-p-Dioxin

1,6-Dichlorodibenzo-p-Dioxin

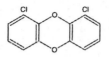

1,9-Dichlorodibenzo-p-Dioxin

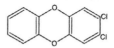

2,3-Dichlorodibenzo-p-Dioxin

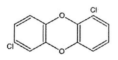

1,7-Dichlorodibenzo-p-Dioxin

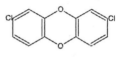

2,8-Dichlorodibenzo-p-Dioxin

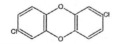

2,7-Dichlorodibenzo-p-Dioxin

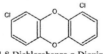

1,8-Dichlorobenzo-p-Dioxin

THE TRICHLOROS

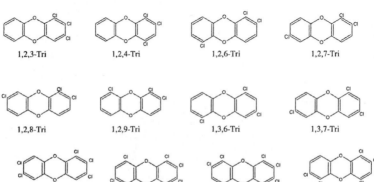

1,2,3-Tri 1,2,4-Tri 1,2,6-Tri 1,2,7-Tri

1,2,8-Tri 1,2,9-Tri 1,3,6-Tri 1,3,7-Tri

1,2,3,7,8-Penta 1,2,3,7,9-Penta 1,2,3,8,9-Penta 1,2,4,6,7-Penta

1,3,8-Tri

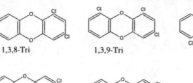

1,3,9-Tri

1,4,6-Tri

1,4,7-Tri

2,3,6-Tri

2,3,7-Tri

THE TETRACHLOROS

1,2,3,4-Tetra

1,2,3,6-Tetra

1,2,3,7-Tetra

1,2,3,8-Tetra

1,2,3,9-Tetra

1,2,4,6-Tetra

1,2,4,7-Tetra

1,2,4,8-Tetra

1,2,4,9-Tetra

1,2,6,7-Tetra

1,2,6,8-Tetra

1,2,6,9-Tetra

1,3,6,8-Tetra

1,3,6,9-Tetra

1,3,7,8-Tetra

1,3,7,9-Tetra

1,4,6,9-Tetra

1,4,7,8-Tetra

2,3,7,8-Tetra

1,2,7,8-Tetra

1,2,8,9-Tetra

1,2,7,9-Tetra

THE PENTACHLOROS

1,2,3,4,6-Penta

1,2,3,6,7-Penta

1,2,3,6,8-Penta

1,2,3,6,9-Penta

1,2,4,6,8-Penta

1,2,4,6,9-Penta

1,2,4,7,8-Penta

1,2,4,7,9-Penta

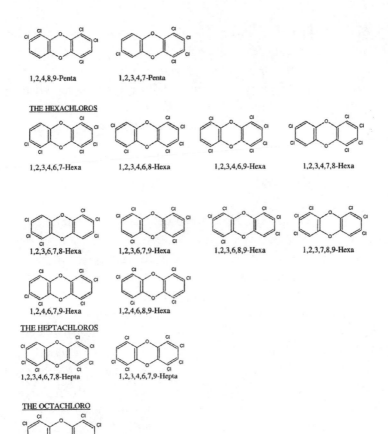

1,2,4,8,9-Penta   1,2,3,4,7-Penta

THE HEXACHLOROS

1,2,3,4,6,7-Hexa   1,2,3,4,6,8-Hexa   1,2,3,4,6,9-Hexa   1,2,3,4,7,8-Hexa

1,2,3,6,7,8-Hexa   1,2,3,6,7,9-Hexa   1,2,3,6,8,9-Hexa   1,2,3,7,8,9-Hexa

1,2,4,6,7,9-Hexa   1,2,4,6,8,9-Hexa

THE HEPTACHLOROS

1,2,3,4,6,7,8-Hepta   1,2,3,4,6,7,9-Hepta

THE OCTACHLORO

1,2,3,4,6,7,8,9-Octa

Dioxin Cousins – The Furans

*"He telleth the number of the stars; He calleth them all by their
names", Psalms 147:4.*

Although furan chemistry is considerably different from dioxin
chemistry, chlorinated furans are very similar to chlorinated dioxins.
There are 135 different chlorinated furans (135 congeners). Each of
these has it's own unique properties and name. Some are almost as
toxic as their comparable dioxins.

As the analytical chemistry of chlorinated dioxins advanced from measurement of 2,3,7,8-tetrachlorodibenzo-p-dioxin to that of all 75 dioxin congers, the chlorinated dibenzofurans were sometimes reported singly, sometimes as a group. Thus data appears which has many different headings making proper interpretation more difficult. The heading CDD/F's means the total of all chlorinated dibenzo-p-dioxins and all chlorinated dibenzo-p-furans.

Other Dioxin Families

Seventy-five bromo (Br) dibenzo-p-dioxins also exist and the structures are written exactly as the chlorinated ones. Other groups, which are known to replace hydrogen on the dibenzo-p-dioxin rings include nitro (NO_2), hydroxyl (OH), methyl (CH_3), and amino (NH_2).

It is possible to have various combinations of these groups on the dibenzo-p-dioxin rings. For example, chloro—and bromo—groups are known to exist together on the same molecule. When this occurs 1700 different molecules are postulated. Since there are also 75 possible chlorodibenzo—and 75 possible bromodibenzo-p-dioxins, a total of 1850 different "dioxin" molecules can possibly exist.

That dioxins constitute a family was acknowledged by the Freshwater Foundation in an article, "Dioxin and Their Cousins: The Dioxin '91 Conference", Health and Environment Digest, 5, 1, 1992.

THE BASKING OF "DIOXIN"

"If there is one way better than another, it is the way of nature",
Aristotle
"The sun discovers atoms and makes them dance naked in his
beams", D. Culverwell.

February 1970

The voice of the Dow manager on the telephone sounded excited. "Our Director of Research is up in the air – somewhere between here and Washington, D.C. When he lands he will call to learn what we know about the half-life of 2,3,7,8-TCDD in a solvent exposed to sunlight. Do we have such data?

"I'm sorry, we don't!"

"Is there some way we can get such data by eleven this morning?"

"There is one possibility!" (I had previously very briefly studied the behavior of a solution of pentachlolophenol exposed to a sunlamp. That compound had been used to treat the Nile River to destroy infestations of slugs. One mile downstream the pentachlorophenol could no longer be detected. The exposure to light from a sunlamp showed that the pentachlorophenol had been destroyed.

I hung up the phone and looked at my watch. It was 9 a.m., February 5, 1970. I rushed down the hall to confer with Dr. Rudy Stehl. Fortunately, I found him at his workbench.

As a part of a different project, Rudy had already prepared solutions of 2,3,7,8-TCDD in methanol and isooctane. Together we devised an experiment whereby these solutions in quartz spectrophotometer cells were irradiated by light from a sunlamp. Degradation was measured by ultraviolet spectrophotometry. By 11:15 a.m., Rudy had already shown that the dioxin was destroyed by light.

When the call came at noon, we were able to report a half-life of three hours. On February 16, this information appeared in Chemical & Engineering News without reference to its source. This was a great shock to us as our study was very preliminary and no funds were available for extending or duplicating our study. Furthermore, the United States Department of Agriculture had started a study which would take six months and would be a complete study. We felt our scientific credibility was in jeopardy. So, we waited with great concern for the USDA results. When they were reported we were thrilled to see that the half-life they reported was within 10 minutes of our report.

Usually scientists submit their work to a scientific journal and the work is subjected to "peer review". That is, other scientists read and criticize the work. Careful "peer review" usually prevents bad data from appearing in print. Unfortunately, much dioxin information was published before it could be confirmed by other scientists. As a result, much confusion was evident in the early work. What was there about dioxin which caused scientists to rush to judgement, jeopardizing their own integrity?

Much later scientists recommended that sunlight be used to destroy dioxin at Seveso, Italy, and Times Beach, Missouri. But political authorities were fearful of using it. So the contaminated soil in Italy was entombed and that at Times Beach incinerated.

CHARACTER MATTERS

"Compounds are not marked by nature with chemical formulas,
but by properties, and it is by these we have to distinguish them",
J. D. Henricks.

All chemical compounds whether isolated from nature, created from raw materials in a laboratory, or isolated from chemical products, behave in a totally consistent manner. Understanding this behavior allows humans to handle the materials with safety. (Indeed, the leading American chemical companies have established one of the best safety records of any industry anywhere).

The character of molecules is determined by measuring their physical and chemical properties. Once known, the behavior of the molecules can be predicted and relied upon. This means that molecules can be controlled and managed. Banning a molecule as a product simply means that those making the decision have no faith in either the behavior of molecules or the integrity of men to manage them. According to measurements made in Dow's laboratories, 2,3,7,8-tetrachlorodibenzo-p-dioxin (TCDD) is a molecule. We know it is a molecule because we have characterized it. This characterization has been confirmed and brilliantly extended by many other investigators in several other laboratories working independently. Such determinations are usually routine, get minimum attention in the scientific literature, and hardly ever mentioned in the lay press.

TCDD is not intentionally manufactured. It occurs as a trace impurity in some chlorinated products such as trichlorophenol. The difficulty of isolating enough for characterization was greater than the problems of synthesis. Therefore, scientists at Dow synthesized several grams. A portion of this synthesized material was used to grow a crystal, which was examined by single crystal X-ray diffraction. This examination alone

established that the molecule is essentially planar and has the molecular structure as shown in the preceding chapter. In addition, the crystal was examined by infrared spectrophotometry mass spectrometry and nuclear magnetic resonance, producing spectra, which were consistent with the structure given. Further, the carbon, hydrogen, and chlorine content was qualitatively and quantitatively correct for this structure. The crystal had a molecular weight of 322 and a melting point of 303-305°C.

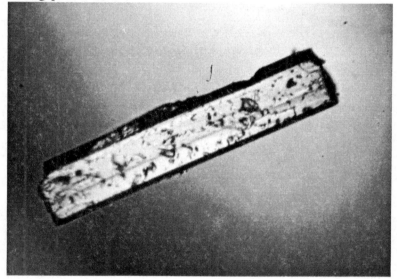

Crystal of 2378-TCDD, magnified 250X under polarized light. Photo by Howard Garrett. Used to determine physical properties.

The weight of a molecule is inconceivably small – only 5.4 X 10^{-22} grams or 0.00000000000000000000054 grams. Water solubility is a behavioral property which has a major influence on environmental fate. Dr. Norman Skelly, of Dow, determined the solubility in various solvents. These are as follows:

SOLUBILITY OF 2378-TCDD IN VARIOUS SOLVENTS

Solvent	Solubility (grams/liter) at 25°C
0-Dichlorobenzene	1.4
Chlorobenzene	0.72
Benzene	0.57
Chloroform	0.37
Acetone	0.11
ā-octanol	0.048
Methanol	0.01
Water	0.0000002 (0.2 ppb)

The determination of the solubility in water is very difficult and other investigators have obtained various unpublished results. Some believe the solubility is higher, some believe lower. Difficulty arises because particles suspended in the water adsorb TCDD causing the solubility to read higher; on the other hand, TCDD is also adsorbed on the walls of the container causing the solubility to be lower. Nevertheless, the solubility is so low compared to that in organic solvents that reasonable predictions can be made about how the molecule will behave in nature.

Because 2,3,7,8-TCDD is believed by many to be much more toxic than other dioxins, some investigators refuse to work with it but use 1,3,6,8-TCDD and OCDD to measure water solubility. Thus the former has been measured at 0.32 part per billion at 20°C and 0.39 part per billion at 40°C while OCDD gives 0.4 part per trillion at 20°C and 2.0 parts per trillion at 40°C.

Another important behavioral property is vapor pressure. 2,3,7,8-TCDD is not very volatile but it does gradually disappear into the air. It's vapor pressure has been reported to be 7.2×10^{-10} atmospheres. Again, most investigators refuse to work with the "toxic" 2,3,7,8-TCDD. In any case, we know that 2,3,7,8-TCDD has some tendency to mix with air because it can be put through a gas chromatograph into a mass spectrometer. When a crystal is sealed in a glass vial, small crystals will appear on the inside walls of the vial after about a year.

Physical constants are useful for the prediction of the behavior of

molecules between various types of matter. The octanol-water partition coefficient helps predict how molecules will be distributed between water and body fats and oils. Henry's constant is useful for
understanding how molecules are distributed between air and water.
Adsorption coefficients are used to predict the distribution of molecules between solid particles and solvents, including water. All of
these can be estimated from data taken on 1,3,6,8-TCDD and
OCDD. These estimates are useful for general predictions but do not
entirely satisfy those who are concerned by factors of ten difference at
these extremely low levels.

2,3,7,8-TCDD is a very stable molecule. Its chemical stability is
remarkable. It is not attacked by any of the usual oxidizing or reduc-
ing agents. It has no functional groups. It is thoroughly stable up to
the temperature of 850°C after which it is both oxidized and reduced
as explained elsewhere. However, it is unstable in ultraviolet light. Its
absorption peak (maximum) at 310 nanometers decreases in intensity rapidly on exposure to sunlight, indicating a rupture of the benzene rings. Attempts to use light to destroy it in the environment
have either been deemed too expensive or too scary by decision makers.

In summary, it is obvious that TCDD reveals its behavioral secrets very reluctantly. However, scientists now know that it is chemi-
cally inert, stable at temperatures up to 850°C, only slightly volatile
at room temperatures, adheres to surfaces (especially particulate matter), concentrates in oils and fats, but disintegrates in sunlight. From
these characteristics we can predict how it will move in the environment.

References

1. G. R. B. Webster, K. J. Friesen, L. P. Sarna, and D. C. G. Muir, "Environmental
   Fate Modelling of Chlorodioxins: Determination of Physical Constants",
   Chemophere, *14*, 609-622 (1985).
2. T. Mill, "Prediction of the Environmental Fate of Tetrachlorodibenzodioxin.
   Dioxins in the Environment", M. A. Kamin and P. W. Rodgers, Editors,
   Hemisphere Publishing Corporation (1985).

DECADE V
FEAR FLARES

FEAR

"We fear things in proportion to our ignorance of them",
Livy, Roman historian (64 or 59 B.C.-A.D. 17).

By now we noted that fear of "dioxin" caused unusual things to happen. The following statement describes the observation.

"Dioxin" is the only molecule the fear of which has ever caused:

1. The National Academy of Sciences to send a special commission to a prestigious university to investigate the integrity of an analytical method. 1973.

2. Twenty-six Italian women, exposed to dioxin in 1976 during the Severso episode, chose to have therapeutic abortions in spite of the vigorous opposition by the Pope. No abnormalities, even in the chromosomes, were found in the aborted fetuses. (Rehder, et. Al., Schweiz. Med. Wocheuschr), 108, 1617, 1978.

3. A state and municipal official to stop the operations of a municipal incinerator. "Dioxin Haunts EPA Refuse Plants," ENR, February 19, 1981.

4. Das Mozarteum-Quartett Salzburg to play a special concert on its behalf. Dienstag, 12. October 1982, 19.00 Uber. Im Rittersaal per Residenz. Festliches Kammerkonzert. 3. Internationaleu Tagung Uber Chlorierte Dioxine.

5. A war veteran's group to blame all their physical maladies on it. Wilcox, F. A., "Waiting for an Army to Die", Vintage Books, New York, 1983.

6. A U. S. federal agency to buy an entire small town. Chicago Tribune, "The Times Beach Buyout", March 2, 1983.

7. U. S. Congressman Scheuer to hold a special congressional hearing, March 22, 1983.

8. A president of a large international company to promise to "leave no stone unturned" in his campaign to search for traces of it. Midland Daily News, June 1, 1983, p. 1. New York Times, June 2, 1983, p. A19.
9. The creation of a billion dollar industry without a product. EPA Scientist, Personal communication. January 24, 1984.
10. A German company to shut down a herbicide manufacturing plant, "Hamburg Faces Dioxin in the Wind", New Scientist, July 26, 1984.
11. Analytical scientists to make quantitative measurements at parts per quadrillion levels in environmental samples. The Dow Chemical Company, "Point Sources and Environmental Levels of 2,3,7,8-TCDD on the Midland Plant Site", November 5, 1984.
12. A state governor to declare a state of emergency in a large city, "Dioxin Discovery spurs emergency state in Paterson", The Harold News, Passaic, NJ, June 13, 1985.
13. The Philharmouische Cellisten Koln to give a concert at the Markgrafliches Operhaus Bayreuth, September 17, 1985, 20 Uhr. Dioxin 85.
14. The Federal Register to reference it 215 times in 5 years.

Consideration of all these unusual and dramatic events leads to a difficulty in believing dioxin is merely a molecule. Its effects on humans are more like those of a mischievous evil spirit. With or without contact it has profound effect on the behavior of human beings.

THE SEARCH FOR ZERO

"I'm an idealist. I don't know where I'm going, but I'm on my way", Carl Sandburg.

1970

In the early 1970s regulators, with the support of many advocates, continuously and vigorously spoke of:

zero-based[11]
zero tolerance[2,5,12]
zero emissions[1]
zero effluents
zero pollution
zero risk[4,7,8,13,16]
zero exposure[5]
zero content
zero radiation[4]
zero discharge
zero defects
zero contamination[4]
zero-level concentration[6]

The media found these terms newsworthy and promoted their use so extensively that it seemed to some of us that a wave of "zero mania"[13] had gripped the nation.

Of course, regulators never meant to have anyone think of these zero terms in a scientific sense. As explained by Sanford A. Miller[3], Director of Foods, U. S. Food and Drug Administration, "Each generation of humankind has had its search for perfection. In the Middle

Ages, reflecting the intense religious nature of medieval society, this search expressed itself in the quest for the Holy Grail. During the Renaissance, with the rise of secular mysticism, it took the form of the search for the philosopher's stone, a device capable of providing absolute understanding and perfect truth to all who owned it. In the age of exploration it was the search for ideal platonic lands, while during the last century, it was the quest for rationality, resulting from the conviction that man's knowledge could encompass all of nature. The search for zero represents, then, a legitimate descendant of the medieval search for the Holy Grail."

But, just as searchers for the Holy Grail did not know precisely what they were looking for, the modern searchers for zero did not know what zero is. They neatly avoided the necessity to know by demanding that trace analysis be done by the "best available technology".[1] This put great stress on analytical scientists who had to decide what is the "best available technology". Should one use methodology that all laboratories can use and whose reliability has been demonstrated over and over? Or should it be newly developed methodology with a significantly lower limit of detection and a higher degree of specificity, but also with a greater amount of uncertainty.

Because molecules of any particular compound are always present in the environment at some concentration level, one analytical method may report that they are found while another may not find them. The situation has been well described by Dr. Gunter Zweig who referred to the phenomenon as the "vanishing zero".[9, 10] Zweig pointed out that the point at which one sees an airplane disappear into the sky is "zero". However, the airplane comes into view again if one looks through binoculars and then again disappears into the sky creating a second "zero". A third "zero" is found by the use of a telescope and on and on it goes, as the means of seeing the airplane becomes more sensitive. Furthermore, the eye, the binoculars, and the telescope each have the ability to estimate the size of the airplane. This estimation becomes less and less certain as the airplane disappears from view. An analytical scientist would call the point at which the size of the airplane can no longer be measured reliably the "limit of quantitation" and the point at which the plane can no longer be seen

with certainty the "limit of detection". Concentrations below the
"limit of quantitation" have been referred to as "not sizeable."[18]

The data that Dr. Zweig used to develop his concept of the "vanishing zero" were taken from the lowest reported limit of detection of
various analytical procedures for residue analysis over a period of years.
Although mostly taken from earlier data, it fits very well with the
limits of detection found for methods used to determine 2,3,7,8-
TCDD. These data are compared in the following table.

| Year | Detection Limit, Picograms/Gram | |
	Zweig's Data	Dioxin Methodology
1940	1,000,000	
1950	1,000,000	
1958	1,000,000	
1960	500,000	
1965	1,000	1,000,000
1970	100	50,000
1975	1	10
1980	—	0.2
1983	—	0.01

Others have referred to this as the "receding zero" or the "shrink-
ing zero".[14]

One author[15] put it well. "We are left with the humbling realization that as our limits of observation are expanded, zero becomes an
increasingly elusive phenomenon", he wrote.

Chemists seeking to measure small numbers of molecules will
continue to develop more sensitive and specific instruments for doing so. Eventually, they will reach the ultimate limit – single molecule detection. And what will that mean in a practical sense? Nothing, of course! But as a point of interest, it may mean that at least one
molecule of every substance that has ever existed in nature will be
present in a glass of drinking water.

Meanwhile bureaucracies have sprung up on a global basis, midst
great publicity, sponsoring seminars and graduate courses on "zero
emissions" and using business men and "educators" more than scien-

tists to promote the concept. With the glory and the money flowing to these people, who will do the monitoring?

References

1. Environmental Defense Fund, Memorandum to EPA proposing zero emissions standards.

2. Department of Labor, OSHA, "Carcinogens", Federal Register, 39, No. 20, January 29, 1974.

3. S. A. Miller, "The Search for Zero", Biomed. Mass Spec., 8, 375 (1981).

4. M. H. Bradley, "Zero-What Does That Mean?", Science, 208, (1980).

5. W. B. Crummett, "The Search for Zero", J. Environ. Sci. Health, (1978).

6. R. P. Mariella, "The Casual Use of the Word Toxic", Chem. & Eng. News, 60, 43, (June 7, 1982).

7. S. Preston, "Toxic Chemical Detection Difficult Task", Bay City Times, Sunday, January 6, 1980.

8. J. M. Callahan, "First U. S. Environmental Chief questions EPA Car Standards", Michigan Living/Motor News, December 1979.

9. G. Zweig, "The Vanishing Zero-Recent Advances in Pesticide Analysis", Essays Toxicol. 2, 156, (1970).

10. G. Zweig, "The Vanishing Zero-Ten Years Later", J. Assoc. Off. Anal. Chem., 61, 229 (1978).

11. F. B. Jueneman, "Arsenic-An Old Case", Ind. Res. Dev., May 1979.

12. R. P. Jones, "A Little Bit of Intelligence", Ind. Res. Dev., November 1978.

13. P. B. Weisz, "Selenium: Another Window to Reality", Chemtech, 13, 45, (1983).

14. W. B. Crummett, T. J. Nestrick, and L. L. Lamparski, "Analytical Methodology for the Determination of PCDDs in Environmental Samples: An Overview and Critique", pp. 57-58. "Dioxins in the Environment", M. A. Kamrin and P. W. Rodgers, editors. Hemisphere Publishing Corp., Washington, 1985.

15. Walrus, "Dioxins, Dioxins Everywhere", Chemistry.

16. D. E. Koshland, Jr., "Immortality and Risk Assessment", Science, 236, 241 (1987).

17. K. Weissermal, "Advance into the Imponderable".

18. J. Angerer, "Expert discussion of the analysis of industrial and environmental

medically relevant chlorinated hydrocarbons in biological materials and air",
Frenzenius Z Anal. Chem., *325*, 351 (1986). Cover letter to M Leng.

19. James H. Krieger, "Zero Emissions Gather Force as a Global Environmental
Concept", C&E News, July 8, 1996.

20. Zero Emissions Research Institute, "The Second Annual World Congress on
Zero Emissions", Chattanooga, TN., U.S.A., May 29-31, 1996.

21. W. E. Harris, "Analyses, Risks, and Authorative Misinformation", Anal. Chem.,
64, 665A, July 1, 1992.

IT'S E-T-C-Y-L

"Molecule! Molecule! Where have you been?
'I've been to Midland to see Etcyl Blair!'
Molecule! Molecule! What did you there?
'I met many scientists who work with great care!'
W. Crummett, 1993.

1970–1988

"This is Etcyl Blair!" "Good morning, Edsel. What can I do for you?" "My name is Etcyl, spelled E-T-C-Y-L! It is not E-D-S-E-L! I am not a flop! I am Etcyl, Etcyl Blair!" So, I recall, went my first professional contact with Dr. Etcyl Blair, a new chemist on the Dow research block. (He wanted to be sure I didn't think of him as the Ford Motor Company's Edsel which did not sell and was referred to as a "flop".) He called because he had prepared some organophosphorothioate compounds whose stability in water needed to be measured and there were no methods for doing this. He wanted to know if I had a reasonable idea on how to do this. I did. Then he at once proved himself to be a man of grit for he delivered the samples together with a work order and only asked how long it would take.

The proposed experiments worked as predicted and thus began an enduring professional relationship. Soon, however, Etcyl became a manager and advanced rapidly through organic synthesis, agricultural chemistry research, and into environmental and health research, becoming a vice president of the company. Meanwhile I remained in analytical chemistry research. Each time Etcyl was promoted he let me know that he now had his own analytical people and didn't need me anymore. Soon, however some extraordinary event would bring him back with a plea for help.

So, on these unusual occasions, I served as Etcyl's "sidekick". In 1970, he received an urgent call to come to Washington, D.C., to consult with U. S. Department of Agriculture scientists. I was recruited late one night and the next day we left early for Washington and visited with the government scientists. We talked all day and got all sorts of hints about future regulatory action. But afterwards, on the plane returning home, we asked each other, in wonderment, "Why were we called to visit?" Neither of us could answer the question. I had a deep feeling of foreboding.

In 1973, I was with Etcyl and the National Academy of Sciences Committee, led by Prof. Anton Lang of Michigan State University, to investigate the integrity of the Harvard University analytical method for determining 2,3,7,8-TCDD in environmental samples. Someone had told me the method produced unequivocal data. Based on calculations by Dr. Lewis Shadoff, I suggested that the data, like all other trace quantity data, was equivocal. The method was approved. (Later, Baughman and Meselson encountered an interference in some samples from Vietnam. Apparently 2,3,7,8-TCDD was produced in the analytical process.) The following day many of the same group met in Washington to consider what data should be collected in Vietnam. A rigorous scientific investigation required more funds than had been appropriated, but no one wanted to ask congress for more. A young scientist suggested grabbing a few samples but the committee refused to be associated with anything other than "rigorous science". Again Etcyl and I were puzzled and wondered what the committee was all about. A well designed rigorous scientific study would probably have relieved many fears – but, if not, the scientific community would have known the truth which was so very important. Why not ask congress to appropriate the money?

I was there, too, when Etcyl informed the Dow production people that they would have to reduce the "dioxin" level in 2,4,5-tricholorophenol and derivatives for political, rather then scientific reasons. I heard their complaints. But they did as Etcyl said and soon met the goal. I know because my laboratory monitored the products.

Etcyl organized the first scientific conference and published (1973) the first book on dioxins, working with the American Chemical Society. Then, with the famous attorney, Milton Wesel, he championed two Dispute Resolution Conferences" on dioxins and related compounds. I was active in all of these endeavors and learned firsthand that environmentalists and regulators were unappreciative and determined to avoid scientific input as much as possible.

Dr. Etcyl Blair recalls our early struggles.

More importantly, within Dow, Etcyl developed a group to teach leaders at every Dow location what environmental and human health responsibilities really meant. The lessons learned from this effort kept Dow as a leader in health responsibilities and were a great lesson for all of us.

Although most Dow managers tried to avoid entanglements with adversaries who challenged the effects of chemicals on society, they did not hesitate to severely criticize Etcyl's handling of these matters. The use of sound scientific data (rather than emotion) for decision making was always under fire, but Etcyl continued to hold firm.

"It's Etcyl calling!" got to be an expected greeting when I answered the telephone. From the first Etcyl was right. He was not a flop!

Etcyl understood that "dioxin" is a molecule! It appeared that some managers and environmentalists did not!

TRACER COSMOS

"We are finding less and less of a whole lot more",
Robert D. Kross and Stanley C. Lewis [3], The Environment
Forum, March 1983.

It is of the utmost importance to recognize that as the concentration level at which the analytical chemist can detect molecules is lowered by an order of magnitude (a factor of 10), the number of compounds that can be detected increases ten times. As an example, tap water treated for domestic consumption contains about 1 part per million of organic compounds. So, such water can be said to be 99.9999 percent free of organic molecules or 99.9999 percent pure with respect to organic compounds. If a complete analysis is made at a detection level of 1 ppm, only one organic compound could possibly be detected; at a detection level of 1 ppb, as many as 1000 compounds may be detected; and at a level of 1 ppt, one million compounds. At the part per quadrillion level of detection, one billion different compounds could be detected if that many were to exist. So far, more than 7 million compounds have been registered by Chemical Abstracts, the abstract journal of the American Chemical Society. These types of considerations led William T. Donaldson of EPA's Athens Research Laboratory to write, "One would expect to find every known compound at a concentration of 1 part per quadrillion or higher in a sample of drinking water."[1]

Very few substances are as pure as drinking water. A purity of 99 percent is more realistic. This case has been thoroughly discussed by Gunnar Widmark of the University of Stockholm, who called the phenomena "Tracer Cosmos".[2]

With 99 percent purity, and at a detection level of 1 ppm, 10,000 compounds may possibly be detected; 1,000,000 at 1 ppb; and

10,000,000,000 at 1 ppt. In the real world it is more likely that about 100 compounds are present at the part per million, or higher, level; 100,000 at the part per billion, or higher level; and so on. The point is that the number of organic compounds which may be reported in a thorough qualitative analysis of a particular sample varies greatly with the concentration level at which the impurity can be detected.

References

1. W. T. Donaldson, "Trace Organics in Water." Environ. Sci. Tech., *11*, 348, 1977.
2. G. Widmark, "Tracer Cosmos, A Realistic Concept in Pollution Analysis", International Symposium on Identification and Measurement of Environmental Pollutants, Ottawa, Canada, pp. 396-398. (1971).

THE HARVARD SURPRISE

"You can always tell a Harvard man, but you can't tell him much", James Barnes.
"If we would become involved with the spirit of the new philosophy of chemistry, we must begin by believing in molecules", J. P. Cooke, Harvard, ca. 1870.

25 September 1970

On September 25, 1970, I stood on the Harvard University campus in awe of the place, and especially its professors. There could be no doubt about that. Society had made certain of it. My "preceptor", Prof. William MacNevin at Ohio State University, being a graduate of Harvard, drilled it into me. Judgments of the media, as well as employers, biased me in my feelings toward the university. I believed truly that Harvard University professors possessed complete knowledge and their integrity in treating such knowledge to be impeccable.

Overwhelmed by tradition and prestige, which I sensed throughout the campus, I entered the chemistry building and asked to see Professor Matthew Meselson. No such professor was known to the chemistry department. I crossed the street to the department of Physical Sciences. No such professor was known to that department. Back in the Chemistry Department, I asked for Robert Baughman, a graduate student, and was told that he worked with a professor in the Biology Department. The several block walk to the Biology Department was rewarding. I learned that the meeting was ready to start in the Faculty Club several blocks beyond the chemistry building. I struggled up the hill, the large briefcase containing all the information I had on the analytical chemistry of 2,3,7,8-TCDD weighing me down. (The confusion in finding Professor Meselson resulted because Bob Baughman's kind invitation with directions dated September 21, 1970, had not reached my office before I left Midland on September 24.)

Breathless I arrived at the top of the hill. I was delighted and relieved to see Dr. David Firestone of the U. S. Food and Drug Administration coming up the other side. Quickly we were joined by Dr. Jack Plimmer and Dr. Edwin Woolson of the U. S. Department of Agriculture. Together we entered the Faculty Club and waited for the appearance of our Harvard professor. Other university professors joined us. Among them were Professor Klaus Bieman, renowned mass spectroscopist from the Massachusetts Institute of Technology, and Professor Riseborough, famed environmental scientist from the University of California. I sat there totally intimidated by my superiors, although I found some comfort in knowing I had interesting information in my briefcase. (I expected to be told we needed a method sensitive to about 1-10 parts per billion.) I was totally unprepared for the question that was raised, however.

Matt Meselson's charismatic personality radiates from a base of superior intelligence and absolute self-confidence. So he was the dominant presence in the room. He explained that the American Association for the Advancement of Science Team had brought back biological samples from Vietnam and that these needed to be analyzed for 2,3,7,8-tetrachloro-dibenzo-p-dioxin at a detection limit of one part per trillion (ppt). The information in my briefcase was useless. I had never even thought about analysis at such low concentration levels. He had calculated, using DDT as a model molecule, the amount of 2378-TCDD likely to be found in Vietnam fish, from the spraying of specified amounts of Agent Orange. From that calculation, the limit of detection needed to measure the amount was clear – one part per trillion! Professor Meselson had demonstrated great faith in the behavior of molecules in the environment and when, years later, the analyses were made, the limit of detection was found to be just right. He had also calculated 1 ppt from the data reported by Dow scientists in a toxicological study of 2,3,7,8-TCDD fed to guinea pigs. (If the toxicological work had used other animals, the need for a 1 part per trillion detection limit would not have been calculated. A method having such a low detection limit may not have been developed, and the rest of this book would not have been written. Moreover the billion dollar dioxin industry would probably not exist. This illustrates the importance of scientists asking the appropriate question.) The question was: Is 2378-TCDD toxic to humans? How toxic? To answer these questions, modern toxicologists would have known to use rats, dose them by spiking their food (gavage would be inappropriate), with smaller amounts of 2378-TCDD.

The dramatic announcement that a method with a detection limit of 1 part per trillion was needed, stunned the group. The best that any one of us could do was 50 parts per billion in extracts from eagles. This was reported by Ed Woolson of the U. S. Department of Agriculture. This meant that the state-of-the-art was at a detection level almost five orders of magnitude higher than Meselson's requirement. Some of us obviously thought that 1 part per trillion could not be achieved. I, for one, wanted desperately to take the position that such a measure was not needed, but I had no supporting data and the right political words would not come. So I did not say much.

Meselson asked the group how a part per trillion could be determined. No one knew. A brainstorming session followed. Out of that came an idea that was eventually used—high resolution mass spectrometry with multiple ion detection. This was first proposed by Klaus Bieman.

Although the initial plan called for Dow to provide analytical chemists to work in Bieman's laboratory, Meselson and Baughman decided to do the work themselves when use of an MS-9 high resolution mass spectrometer became available to them. Three years later they reported the results of that work.

On September 17, 1971, in an American Chemical Society Symposium in Washington, D.C., "Bob" Baughman reported the progress being made[1.] Then on April 2, 1973, during breakfast at the Governors Inn, Research Triangle Park, North Carolina, a beaming Matt Meselson handed me a copy[2] of a report claiming the reliable detection of TCDD in animal tissues at levels approaching one part per trillion. I read this manuscript while we sat there. There was no doubt in my mind that this was a major advance in the field of analytical chemistry. The method featured the novel use of a multichannel analyzer to produce a stronger signal and synthetic ^{37}Cl TCDD, spiked in the sample before clean-up, to measure recovery of natural TCDD.

References

1. R. Baughman and M. Meselson, "An Improved Analysis for Tetrachlorodibenzo-p-dioxin". "Chlorodioxins-Origin and Fate", Etcyl H. Blair, ed., ACS Series 120, American Chemcial Society, Washington, 1973, p. 92.
2. R. Baughman, and M. Meselson, "An Analytical Method for Detection TCDD (Dioxin): Levels of TCDD in Samples from Vietnam". Environmental Health Perspectives No. 5, 27 (1973).

A BODACIOUS EXPERIMENT

". . . every experiment is like a weapon which must be used accord-
ing to its own peculiar power, as a spear to thrust, a club to
strike: so it is with experiments", Paracelsus.

25 November 1970

"Professor Bo Holmstedt of the Swedish Medical Research Council
is here. He claims that 2,3,7,8-TCDD will pass from a gas chromatography through a jet separator into a mass spectrometer. Would
you like to come to the Biochemical Research Laboratory to meet
him? You must come immediately, as he is leaving for the airport in
30 minutes."

I rushed to meet this genius. For about two years we had attempted to use gas chromatography/mass spectrometry to measure
2,3,7,8-TCDD in environmental samples. But no matter what we
tried, the TCDD was always lost between the gas chromatographic
column and the mass spectrometer.

In ten minutes, the amiable Bo convinced me that 2,3,7,8-TCDD
molecules really could go from a gas chromatograph through a jet
separator into a mass spectrometer. Furthermore, the appropriate instrument was commercially available. It was called the LKB 9000
and was manufactured by an instrument company in Sweden.

Consultation with Dow mass spectroscopists strengthened my
belief that Bo's system would indeed allow 2,3,7,8-TCDD to pass
through a jet separator. They had already had discussions with one of
our consultants, Dr. Jack Holland of Michigan State University. An
LKB 9000 instrument was located at Michigan State University in
the charge of Professor Charles Sweely, with whom Dr. Jack Holland,
a Dow consultant, was associated. We arranged for three of our top

analytical scientists (Lew Shadoff, Rudy Stehl, and Vic Caldecourt) to try an experiment on the MSU equipment. This was done on November 25, 1970. In one hour, enough data were obtained to convince us that the separator did work, allowing about one half of the dioxin leaving the column to enter the mass spectrometer. This, then, was a major historical scientific event!

It was necessary for us to have this equipment if we were to continue developing methods for the determination of dioxins in environmental samples. However, it cost $75,000, which was at least five times more than the cost of any instrument as yet purchased for the Analytical Laboratories. To obtain permission to purchase the instrument, I went directly to the company's Director of Research, Dr. Julius Johnson.

"I was wondering when someone would go that way", Dr. Johnson told me. With his support, the equipment was quickly purchased and we were started on the development of the analytical methodology for the determination of 2,3,7,8-TCDD in environmental samples. To maximize the speed at which we progressed, we brought L. A. "Lew" Shadoff into the group. He worked with R. A. "Dick" Hummel to develop the first methods.

Later I was surprised to learn that Bo Holmstedt had not actually put TCDD through a jet separator and the experiment at the Michigan State was indeed a first.

THE BROUHAHA

"With fact he mixes fiction", Horace,
De Arts Poetica (ca. 20 B.C.).

1971

Along with their announcement of a new analytical method capable of a detection limit of 1 part per trillion, Baughman and Meselson reported finding part per trillion levels of TCDD in fish from Vietnam. This led them to conclude that dioxin "may have accumulated to biologically significant levels in food chains in some areas of South Vietnam exposed to herbicide spraying." This reasonable conclusion was immediately blown out of proportion by the following newspaper account which focused on defoliation by spraying Agent Orange in South Vietnam. "Scientists Bare Perilous Chemical in Vietnam Defoliant", Morton Mintz, The Washington Post, April 6, 1973.

Although this article contained some facts of scientific merit, it set the tone for future articles by ignoring the work and opinions of scientists who had done research on dioxin and gave great coverage to two self-appointed instant experts who had never worked with dioxin. Within three years, a series of sensational articles appeared. They varied considerably in content, but had several things in common. These are: (1) a frightening headline, (2) a scary picture, (3) some valid scientific information, (4) many human interest anecdotal accounts, (5) criticism of government agencies, (6) criticism of Dow, and (7) reference to the opinions of self-appointed experts who had never done research on TCDD. These articles include:

"Spraying Dangers in the Air", Daniel Zwerdling, The Washington Post, January 25, 1976, shows an old fashioned spray gun firing spray toward the reader. It referred to two self-appointed science experts.

"World's Most Toxic Chemical. TCDD: A Manmade Killer Poisoning Environment", Paul Jacobs, The Indianapolis Star, September 5, 1976, shows an airplane spraying a food crop. It doesn't refer to any specific scientist, but does its own speculation and attributes the conclusion to "scientists".

Apparently, Jacob's effort was considered very successful for he was quickly asked by Newsday to prepare a series of four articles exclusively for them. These articles, published in August 1976, depend for effectiveness on anecdotes and Jacob's own interpretations of the "facts", are:

- "TCDD: Deadly Poison in the Food Chain."
- "TCDD: A Chemical Nightmare. First, the Animals Began Dying."
- "TCDD: A Chemical Nightmare. Disposing of a Lethal Vietnam Legacy."
- "TCDD: A Chemical Nightmare. Ignorance Reigns, Industry Thrives."

Other articles appeared as an outgrowth of this work, producing sensational headlines in many newspapers. For example:

- "TCDD's Long, Deadly Record Finally Stirs Alarm", Paul Jacobs, Detroit News, September 9, 1976.
- "Dioxin: To Spray or Not to Spray", Daniel Yost, Northwest Magazine, May 22, 1977, shows a drawing of a dead bird and depends largely on anecdotal stories.
- "Give Us Each Day our Daily Dose", Jerry Shields, Gallery, December 1979, gives anecdotal information great credibility. A picture of loaves of bread being injected with a hypodermic needle labeled "HERBICIDE" with a skull and crossbones is included.
- "The Spraying of Oregon", The Williamette Week, Decem-

ber 1979, depends largely on anecdotes and opinions, including those of self-appointed experts.

Special anecdotal stories appeared. A few of the many examples are:

- "Dioxin Contamination Feared in Home Water", The Oregonian, October 6, 1977.
- "A Cancer-Doomed Vet Blames Viet Defoliant", John Hamill,, New York Daily News, June 11, 1978.
- "Herbicide Link Alleged. EPA to Study Miscarriages", Lorraine Ruff, the Oregonian, July 18, 1978.
- "Dioxin Probe Continues on Farm", Maggie Menard, The Milwaukee Sentinel, July 11, 1978.

Scientists with a great deal of experience working with herbicides and dioxins repeatedly attempt to set the record straight. First, by a new approach – dispute resolution conferences. Two were held as follows:

- "The Rule of Reason Workshop on 2,4,5-T", Washington, D.C., March 8-9, 1973.
- "Dispute Resolution Conference on 2,4,5-T," American Farm Bureau Federation, Arlington, VA, June 4-5, 1979.

Although scientists reached general agreement on most issues during the conference, those scientists that were not personally doing research continued to make sensational statements to the media.

A second attempt to educate the public was talking directly to the press. An example is: "MSU Prof. Says Dow Herbicide Not Dangerous", Fred E. Garrett, The Saginaw News, February 19, 1976.

Unfortunately, such efforts had little effect as articles such as these also appeared: "Family Doctor Says Weed Killer More Perilous Than Thalidomide", The Globe & Mail, February 18, 1976.

Various groups brought scientists together in panels to inform the public. For example, officials of Lincoln County, Oregon, convened a panel to explain the 2,4,5-T controversy to area residents.

The scientists were sharply divided over the issues. In an article titled, "Experts Debate Dioxin for Public at Science Center", Capital Press, Salem, Oregon, August 12, 1977, the following positions were reported. Those scientists who believe that dioxin is so toxic and dangerous, the products containing minute traces of it should be completely banned included:

- A professor of genetics,
- A physician, and
- An aquatic biologist, having a total of 3 years research related to 2,4,5-T and its contamination.

If applied correctly, the minute amounts of dioxin in the herbicide do not pose a health hazard, according to:

- Two professors of agricultural chemistry,
- A professor of plant physiology, and two researchers with the Oregon Department of Agriculture, having a total of 35 years experience.

The different opinions were not very reassuring to the public. However, the differences were to be expected since five of the eight panelists had no research experience with dioxin or 2,4,5-T. Their comments were speculations based on others' work, most of which had not been published.

Scientists continued to speculate on the effects of TCDD and make sensational statements about it. As a result, headlines such as: "Scientists Term Dioxin 'Most Toxic Chemical'", Paul Pintovich, The Oregonian, December 9, 1976, and "Dioxins High Toxic, Insidious", Thomas K. Rohrer, National Resources Register, May 1982. Certainly, dioxins are toxic, but they are not insidious. Their behavior is completely predictable.

Sensational stories continued to appear. "Across America, Dioxin", Lew Regenstein, The New York Times, March 8, 1983, shows a broken doll (or is it a deformed baby?) and a dead fish. As in earlier articles, the E.P.A. is reprimanded. Chemical companies are vilified. The press still likes to consort with the "instant expert", who is still the most vocal scientist in the media.

In addition to the instant experts and sensational stories keeping the dioxin issue alive, the behavior of scientists and government agen-

110 WARREN B. CRUMMETT

cies fuel the issue. When scientists insist on laboratories called clean rooms which cannot be entered or left without a change of clothing, the scary image is enhanced. Also, when samples are taken of soil from the end of downspouts in a residual neighborhood, by humans "wearing hooded white bodysuits, yellow rubber boots, black rubber gloves, and gas masks", people naturally think dioxins are insidious. "Hunting for Dioxin Exposes Field Teams to Array of Hazards", Ron Winslow, The Wall Street Journal, July 26, 1983, gives us a vivid description.

The conclusions of international scientific conferences generally had little effect on sensational stories. However, one at Michigan State University to which the press was invited to participate resulted in more balanced reporting. A few headlines which resulted are:

- "Public Overreacts to Dioxin Threat, Researchers Say", Midland Daily News, December 7, 1983.
- "Fear of Dioxin is Overblown, Scientists Say", David Sedgwick, The Saginaw News, December 10, 1983.
- "Scientists: Put Dioxins in the Right Proportion", Daily Press, Escanaba, Michigan, December 7, 1983.
- "Overrated. Some Scientists Suggest Dioxin Blown Out of Proportion as a Pollutant", Journal, Flint, Michigan, December 7, 1983.
- "Scientists Say Dioxin Isn't a High Priority", Michael McKesson, Midland Daily News, December 10, 1983.
- "DIOXIN: After Years of Worry, Findings are Less Dramatic Than Feared", Hugh McCann, Detroit News, December 12, 1983.

This conference resulted in a temporary lull in news stories about dioxins. However, they are back and continue to this day. To create a sensational story, all an environmental group need do is to come to an industrial community and make a ridiculous demand. Even while I write this, Greenpeace is in town demanding that "Michigan prohibit Dow from discharging dioxin from the Michigan Division at any level". "When they say it is impossible to reduce dioxin emissions to zero, they are saying they refuse to undertake the necessary steps. We can do a lot of things on this planet more difficult than eliminating dioxins", the spokesman, Jay Palter, said. "Group's Report Condemns Dow Practices," Jeff Green, Midland Daily News, July 19, 1989.

Many of the scientists who were instant experts and created sensational stories were dealt a severe blow by D. Merlin Nunn, Nova Scotia Supreme Court Justice. The story was reported "Dioxin Trial Judge Says Witnesses Biased", Pete Earley, The Washington Post, October 6, 1983.

"While I do not doubt the zeal of many of the plaintiff's scientific witnesses or their ability, some seemed at many times to be protagonists defending a position, thereby losing some of their objectivity", Judge Nunn wrote. "There was a noticeable selection of studies which supported their view and refusal to accept any criticism or contrary studies. In my view a true scientific approach does not permit such self-serving selectivity . . . ."

Still today these same scientists contribute to sensational stories.

THE METAMORPHOSIS

"A change came o'er the spirit of my dream",
The Dream – Lord Byron

1970-1988

Suddenly I was in the spotlight. No longer was I considered to be just "a pair of hands" or someone's "right hand man". I no longer belonged solely to the manufacturing process but to Dow's corporate operation as well. I was no longer expected to remain silent but was expected to present new extraordinary analytical technology applied to a wide variety of circumstances. From very little travel experience, I was to travel frequently. From almost no interaction with my analytical chemist peers in government, industry, and university to unprecedented interaction. I was viewed in a new light, both inside and outside of Dow.

Unfortunately analytical chemists in other companies were kept in their laboratories and my attempts to exchange information with them was severely limited. Because they, as I, had been kept out of the public arena, we were strangers to each other. I soon learned that they were suspicious of my motives.

Nevertheless, after reporting to the Office of Science and Technology, Dow Agricultural Chemistry management thought it proper to inform our competition in the manufacture of 2,4,5-T. In February 1970, I attended a meeting with our competitors at the Midland Country Club at which Dow's experience and scientific data were reviewed. Emphasis was placed on measurement techniques including analytical chemistry. The competitors promised to get back to us if they had anything to share or need more know-how from us. We agreed that their analytical chemists should meet with me to exchange analytical methodology and experience.

I was instructed by my management to meet with analytical chemists from other 2,4,5-T producers on May 18, 1970, in Washington, D.C. I was prepared to discuss everything we knew about the analytical chemistry of 2,4,5-T and the determination of 2,3,7,8-TCDD in particular. I was greatly disappointed. Only representatives from Hooker and Monsanto joined me. The Hooker representative, Lewis E. Tufts, was a quality assurance manager. He knew nothing of analytical method details but maintained that Hooker had a purer product and measured 2,3,7,8-TCDD at levels well below that of Dow. He had no need for our methods. Monsanto, on the other hand, sent a newly hired analytical chemist. He had no knowledge of dioxin or the analytical methodology used by Monsanto. He had been instructed to go to the meeting to see what Dow was "up to". I felt bewitched, bewildered, and betrayed.

The following day we met industry representatives in a joint session with the U. S. Food and Drug Administration (FDA) and the U. S. Department of Agriculture (USDA). We had only Dow information to report. The FDA and USDA representatives, however, had no difficulty telling us their experiences in determining various dioxins in chickens, bald eagles, and other matrices. I was amazed to learn that they had isolated enough hexachloro-dibenzo-p-dioxin from chicken fat, to form a crystal and identify it unequivocally by single crystal x-ray diffraction.

These early experiences convinced me that working with hostile competitors was of little value and largely a waste of time. I was still confused. I did not know if the competition was so far ahead of Dow that they saw no advantage in sharing with us or if they thought what we were doing was unnecessary.

Apparently the word was out, however, that Dow had developed analytical methodology of merit for about a month later I had a visit from Dr. Horst Vogel of Boehringer Soh, Ingëlheim, Germany. He requested the Dow methodology for use in a European 2,4,6-T producer's round-robin study of a standard method for determining

2,3,7,8-TCD. This work was done and the results published a few years later. (Dow was the only United States producer to participate in the study.)

On September 16, 1971, Dr. Rudy Stehl of our laboratory reported on the determination of dioxins in chlorinated phenols and related materials at the 162nd National Meeting of The American Chemical Society in Washington, D.C. His was paper No. 81 in the Division of Pesticide Chemistry.

On January 28, 1974, as a part of a Dow team, I met with representatives of the few remaining 2,4,5-T producers in Chicago. There we reviewed Dow methods of monitoring production plants for 2,3,7,8-TCDD and explained our self-imposed plant regulations and product specifications. We offered to work with them if they needed help to regulate themselves in a similar manner. We had no follow-up from them. Again, I was not certain if their technology was better than ours, or if they thought we were only trying to increase their costs and make them less competitive. In any case, this ended my scheduled interaction with 2,4,5-T producers.

Technology exchange with our competitors was difficult and disappointing. Equally difficult, interaction with scientist peers in both government and academic laboratories made our efforts to clean up misunderstandings almost impossible.

Meanwhile, in 1970, Dow reorganized its analytical facilities at the Midland site and I was named Technical Director of the Analytical Laboratories. Among the new policies which we instituted was the concept that although no work would be done for the sake of publication alone, all work would be done of the highest quality, the report written for publication, and the argument made that it should be published. All work relating to matters of the environment and health, we contended, was of little value unless it was published. So we especially became aggressive in publishing work on dioxin, as soon as it had been developed enough that we understood its meaning.

I was thrust into a whirlpool of intense ecological fascination and propelled with unprecedented force into uncharted waters of environmental dread disguised as science.

Getting our papers accepted for publication was another matter.

Industrial scientists had not published extensively and had little cred-ibility among academic scientists who did the peer reviews. So our manuscripts were rejected, usually for political reasons. For example, a paper submitted in 1974, to the journal *Analytical Chemistry* was rejected by the editor because a reviewer pointed out that we had not included data taken by the U. S. Environmental Protection Agency (EPA), even though we had the EPA data under a secrecy agreement and the EPA would not give us permission to publish it. The paper was subsequently divided into three separate papers and published in the Bulletin of Environmental Contamination & Toxicology, where they appeared in 1977 and 1978.[1,2,3] Such delaying tactics continu-ously kept us on the defensive until after 1978 when the managing editor of the journal *Analytical Chemistry* asked us to please give them first chance to publish our dioxin work.

We had attempted to overcome our publishing difficulties by speaking whenever the opportunity arose. These were rare, but in 1973 an invitation from John A. Moore of the National Institute of Environmental Health Sciences (NIEHS) gave us 20 minutes to re-view the extensive analytical chemistry research which Dow had ac-complished.[4] The conference was held at the Governor's Inn, Re-search Triangle Park, North Carolina on April 2-3, 1973. We had much to say but very little time in which to say it.

Government scientists appeared to enjoy sharing information with us. Those in USDA and FDA freely discussed dioxin chemistry. Those in EPA, however, became hostile as soon as the possibility of 2,4,5-T hearings were announced. For example, they refused to eat lunch with us at the conference held at the Governor's Inn on April 2, 1973. They were not allowed by their administrators to visit with Dow scientists.

In 1972, I was on the "horns of a dilemma". Greatly discouraged and saddened by my inability to communicate our scientific findings and adequately express my understanding of the meaning of these findings, I felt the need to withdraw from the debate and bury my-self in technology in some remote laboratory. On the other hand, it appeared obvious that the role of analytical chemists and the mean-ing of their data were so little understood that they needed champions. The champions should come from the academic community. Regrettably financial support for academic analytical scientists seldom came from industry and the interest of these scientists was directed elsewhere. Did I dare to attempt to explain the untenable positions analytical chemists in industry found themselves? Could it be done without appearing arrogant and self-serving?

My opportunities came in 1978. "The Problem of Measurements Near the Limit of Detection"[5], given at a New York Academy of Sciences meeting in New York on June 24, 1978; "The Search for Zero"[6], given at an Engineering Foundation Conference in Rindge, New Hampshire on July 10, 1978; and "Fundamental Problems Related to Validation of Analytical Data"[7], given at the Ninth Annual Symposium on the "Analytical Chemistry of Pollutants", Jekyll Island, Georgia, on May 8, 1979, were all well received by the scientists and seemed to open doors to further interactions in the scientific community.

Thereafter invitations to speak came regularly and the exchange of analytical information was made easy with all except former producers of 2,4,5-T.

References

1. N. H. Mahle, H. S. Higgins, and M. E. Getzendauer (1977). "Search for the Presence of 2,3,7,8-Tetrachlorodibenzo-p-dioxin in Bovine Milk." Bull. Environ. Contam. & Tox., *18*, 123.
2. L. A. Shadoff, R. H. Hummel, and L. Lamparski (1977). "A Search for 2,3,7,8-Tetrachlorodibenzo-p-dioxin (TCDD) in an Environment Exposed Annually to 2,4,5-Trichlorophenoxyacetic Acid Ester (2,4,5-T) Herbicides." Bull. Environ. Contam. & Tox., *18*, 478.
3. C. W. Kocher, N. H. Mahle, R. A. Hummel, and L. A. Shadoff (1978). "A Search for the Presence of 2,3,7,8-Tetrachlorodibenzo-p-dioxin in Beef Fat." Bull. Environ. Contam. & Tox., *19*, 229.
4. W. B. Crummett and R. H. Stehl (1973). "Determination of Chlorinated Dibenzo-p-dioxins and Dibenzofurans in Various Materials." Environ. Contam. & Tox., *19*, 229.
5. W. B. Crummett (1979). "The Problem of Measurement Near the Limit of

Detection." Annals of the New York Academy of Sciences, *320*, 43-47.
6.W. B. Crummett (1979). "The Search for Zero." J. Environmental Science and Health, *14*, 19-34.
7.W. B. Crummett (1979). "Fundamental Problems Related to Validation of Analytical Data." Toxicol. Environ. Chem. Rev., *3*, 61-71.

PERCEPTION & REALITY

"New roads : new ruts." Horace Walpole, 1712-1797.

1972

By now analytical scientists at Dow had good reason to believe that questions about "dioxin" (2,3,7,8-dibenzo-p-dioxin) had been appropriately addressed. This molecule had been identified as an unwanted trace contaminant in 2,4,5-trichlorophenol and its derivatives; synthesized and analytical standards prepared; characterized by all the known measurement techniques; and controlled in Dow products by newly developed analytical methods of confirmed integrity at levels 10 times lower than deemed necessary by Dow toxicologists. Thus we had not only fulfilled all the requirements of government agencies we had gone the extra mile. No other product contaminant had been treated so rigorously. There was no doubt that Dow products were safe when used as directed. Once again we had proven that molecules could be successfully managed. Although there was already some noise decrying this conclusion, we were totally unprepared to weather the brouhaha which developed. "Dioxin" appeared to be a molecule that created fear and caused people to do strange (unscientific) things. Was it more than just a molecule?

We now turned our attention to measuring the amount of "dioxin" in the environment. We were well prepared to develop methodology to do this. We had state-of-the-art equipment and some of the best chromatographers and mass specroscopists in the world. We were confident we could find and measure "dioxin" (if present) in any matrix of interest at appropriate detection levels. Then we could determine how the dioxin got there and discover ways to completely control and manage the presence and movement of this molecule in the environment. This was our perception!

This perception was reinforced by the merger of Dow analytical laboratories. For the first time powerful measurement tools would be under the same management as the separation systems. This created many opportunities for spectroscopists, chromatrographers and instrument development folks to do extraordinary things together. The resulting laboratory, consisting of almost 200 technical people, was managed by a technical director and a technical manager. I was named technical manager, a position without a job description. We were not, however, positioned to communicate effectively with the world outside Dow. Those of us working in trace analysis had neither published much nor attended many scientific meetings outside the company. We had not honed either our writing or speaking skills. So our papers were rejected and we were seldom heard speaking. This took some years to correct as we didn't immediately recognize that we needed to have credibility with our peers in academia, government an other industries. In spite of this naivety on my part, I soon found myself serving on committees, task forces and study groups both within Dow and with government agencies in the United States and Canada.

Immediately it became apparent to me that the integrity of analytical data at trace levels was a major problem. Different laboratories used different principles in the interpretation of signals. Consensus on data interpretation needed to be reached. But of even greater importance was the need to communicate with journalists.

Each encounter with peers and others interested in "dioxin" became a great adventure filled with fun. Often these meetings were humorous, always challenging, and sometimes testy. Scientists from industry were usually portrayed as biased and even incompetent. So "dioxin" appeared to cause people to do strange as well as extraordinary things. By means of anecdotal stories and essays this book attempts to reveal the astonishing truth of the frustrations, pains and satisfactions experienced by this industrial analytical chemist.

DECADE VI
SCIENCE ROARS

A MOST UNFORGETTABLE SCIENTIST

"A man must not swallow more beliefs than he can digest",
Havelock Ellis, The Dance of Life.

1973

I have been privileged to interact with many scientists concerning environmental chemistry and related matters. One of the most unforgettable was Dr. Leonard Axelrod. As a scientist in EPA's Criteria and Evaluation Division, Axelrod played a key role in devising the EPA Dioxin Implementation Plan (DIP). In that plan, Dr Axelrod succeeded in bringing together many divergent viewpoints and meld them into a designed experiment which, if followed to its logical conclusion, would have resolved most of the issues related to TCDD.

Dr. Alexrod had a special insight into matters related to criteria and evaluation. For example, in a pep talk to members of DIP he once said, "If you use a method for the determination of dioxins with a detection limit of one part per billion, I will believe your data at one part per million. Similarly, a detection limit of one part per trillion will produce believable data at one part per billion and so on". At the time (1973) I thought him to be more than a bit conservative. However, more experience in analysis at these very low concentration levels has convinced me that he was far wiser than the rest of us at that time.

In a private conversation he once said to me, "I will put science into this agency (EPA) or die trying!" Six months later he died of a heart attack. Others, equally gifted, qualified, and committed are still trying, and society is beginning to experience the positive results of that effort. The benefits are too great to be measured.

After 25 years of picking and choosing data to support the regulation of "dioxin", the EPA has been chided by its own Science Advisory Board for failing to keep politics out of the scientific process. Perhaps science will eventually be allowed to play its rightful role.

Reference

1.Katheryn E. Kelly, "Cleaning Up EPA's Dioxin Mess", The Wall Street Journal, June 29, 1995.

EPA DIOXIN
IMPLEMENTATION PLAN

"Unless we can really measure it, we know nothing about it",
F. D. Rossini, paraphrasing Lord Kelvin.
"No one wants advice – only corroboration", John Steinbeck,
1902-1968.

25 July 1974 – 1977

"Where is Lew Shadoff?", EPA's Gunter Zweig greeted me in the lobby of the Mayflower Hotel, Washington, D.C., on July 25, 1974. "I was sure you'd bring him. We want to talk specifically about mass spectrometry and I have evidence to suggest that he is one of the very best mass spectroscopists."

"I'm sorry! I thought this meeting was called to decide whether EPA and Dow should jointly investigate dioxin in the environment." "Yes, but we want to specifically talk about mass spectrometry and need Lew."

"If you want Lew you will have him," I said thinking fast. (Since the conference was scheduled for two days, I expected that I could have Lew come for the July 26 session.)

I called Lew at about 9 a.m. and asked him to come "as soon as he could get a plane". To the amazement of everyone, he arrived shortly after lunch. "It would have taken at least two weeks for EPA to get a scientist here", Gunter opined.

Lew arrived without luggage or toothbrush. Neither did he have any research notes or visual aids of any kind. Worse yet, he had not found time to inform his wife. Nevertheless, he was quickly asked to lecture on the use of mass spectrometry for the determination of chlorinated dioxins in environmental samples. His ad lib technical talk drew high praise from EPA mass spectroscopists and it was decided to use Dow analytical methodology for the study. To support this effort we would have to be leaders. It was a scary thought.

Dr. Lewis Shadoff ad-libs beautifully.

The spirit with which Lew responded to the challenge of dioxin determination was contagious and had a great influence on the reaction and behavior of EPA scientists as well as those in academia. Quickly, the most ambitious and comprehensive analytical plan ever devised was formulated. Called the Dioxin Implementation Plan (DIP), a protocol was designed to optimize monitoring and analytical research. Problems of sample selection, sample contamination, background noise, interferences, signal detection, signal measurement, identification of cause of signals, and confirmation of data were to be appropriately addressed. In other words, the analytical methodology was to be simultaneously developed, applied, validated, and confirmed. The resulting data were to be interpreted with the agreement of all participating laboratories.

To oversee the collection and analysis of samples, a team consisting of representatives from the U. S. Environmental Protection Agency, the U. S. Department of Agriculture, the Environmental Defense Fund, and Dow was organized. The analytical measurements were made by the EPA's Pesticides and Toxic substances Effects Laboratory (PTSEL), Harvard University, Dow Chemical Company, and EPA's contract laboratories. Initially, the EPA contract laboratories were Wright State University and the University of Utah. Later, the University of Nebraska (Lincoln) replaced Utah.

Scientists from the laboratories doing the analytical work were called the Dioxin Analytical Collaborators. These changed to some extent with time. Members of the DIP team were:

Wright State University	Thomas O. Tiernan, Michael Taylor
U. S. Environmental Protection Agency	Ralph T. Ross, Edward O. Oswald Aubry Dupuy, Jr., Han Tai, Robert L. Harless
U. S. Department of Agriculture	John Spaulding, W. Hearon Buttrill
University of Utah	Jean Futrell, Thomas A. Elwood
University of Nebraska	Michael Gross
Harvard University	Matthew Meselson, Robert Baughman Patrick O'Keefe
Environmental Defense Fund (EDF)	Maureen Hinkle

The Dow Chemical Company   Warren B.Crummett
                                  Richard A. Hummel
                                  Lewis Shadoff,
                                  Rudolph Stehl

Ably lead by Ralph Ross, later by W. Thomas Holloway and Richard Troast, and still later by Carolyn K. Offutt, the DIP Analytical Collaborators were making rapid progress and were approaching agreement when in 1977, the EPA appeared to lose interest and opted to pursue the matter in legal instead of scientific arenas. This action appeared to be in direct response to Maureen Hinkle's statement that science had failed and EDF would pursue its goals in court.

Scientific progress had not come easily. Initial results were disappointing, indicating the complexity of the problem. The data showed that more than half of the dioxin added to blank beef fat was lost in the preparation of the sample for measurement. This could lead to results which appear to be negative when, in fact, dioxin was present. Such results are called, "false negative".

On the other hand, unexpected interferences were observed. These appeared to be other chlorinated compounds that were not completely removed by the clean-up procedures. Such interferences could be measured as dioxin when in fact they were not present at all. Such results are called, "false positive".

To truly measure the dioxin content of an environmental sample, both "false negative" and "false positive" results must be avoided. It, therefore, becomes necessary to use extensive procedures to remove the interferences without losing any of the dioxin. This is done by a series of extractions and chromatographic separations. The Dioxin Analytical Collaborators attempted to keep the separation of dioxin from interferences constant by having all these separations done by the EPA Bay St. Louis Test Facility under the direction of Dr. Han Tai. Thus, the only variable was in the gas chromatography/mass spectrometers used by the various laboratories. So, in effect, we were really studying the relative effectiveness of the various gas chromatography/mass spectrometry systems.

Even so, the use of gas chromatography/high resolution mass spectrometry for analysis at the 10 part per trillion level is so difficult that a great deal of collaborative effort was required before any two collaborators obtained consistent data. Part of the difficulty arose because different laboratories used different criteria for the identification and measurement of dioxin. The results improved greatly after some criteria were agreed upon. After lengthy debate, the following were developed:

1. Signal should be at least 2.5 times "noise" measured peak to peak to be reported. Otherwise, N.D. (not detected) should be reported.
2. Samples giving a signal between 2.5 and 10 times noise will be reanalyzed whenever possible.
3. To be considered unquestionable, the signal must be at least 10 times noise and the isotopic ratio of 322 and 320 atomic mass units must be correct.
4. Recoveries must be at least 50 percent to be reported. The signal to noise ratio of 2.5 to 1 was called the "limit of detection" and that of 10 to 1 the "limit of suspicion". The term "limit of suspicion" was suggested by Dr. William Upholt, and means the signal above which there is no doubt the analyte is present and can be measured.

Later, Robert Harless and other EPA scientists (Harless, 1980) published a set of criteria which has become the standard for mass spectroscopists making dioxin determinations.

The Dioxin Implementation Plan was designed to determine the 2,3,7.8-TCDD content of beef fat and liver from range land in Missouri, Oklahoma, Kansas, and Texas. The samples were sent to the EPA Bay St. Louis, Mississippi Test Facility where the fat was rendered and the dioxin extracted. The extracts were then cleaned up and sent to the laboratories of EPA at Research Triangle Park, NC, Harvard University, University of Utah, Wright State University, and Dow Chemical Company, Midland, MI for analysis by gas chromatography/mass spectrometry.

On June 16, 1976, the Collaborators met in Washington, D.C. to discuss these results. Final data were available from Wright State University, Harvard University, and Dow. The Collaborators concluded:

1. Of the 85 beef samples analyzed, one shows a positive 2,3,7,8-TCDD level at 60 ppt; two samples appear to have levels at about 20 ppt; five may have levels which range from 5-10 ppt;
2. The analytical method is not valid below 10 ppt;
3. Of the 43 liver samples analyzed, only one suggests any TCDD residue, but the signal was too close to the limit of detection to be measured and a fat sample from the same animal showed no TCDD. Three of the liver samples from cattle whose fat had shown positive TCDD results, showed no TCDD residues.

The EPA had recorded the state, the farm, the age and physical condition of the cow, the application rate of 2,4,5-T herbicide, and the grazing time. This provided a database from which various hypotheses could be formulated. The collaborators were unable to reach a consensus on a specific conclusion.

The Dow data taken on the 85 samples of beef fat showed only eight reportable numbers: one positive result at 70 ppt (signal to noise ratio greater than 10); six possibly positive results at about 20 ppt (signal to noise ratio equal to 5 to 7); and one possibly positive result at about 10 ppt (signal to noise ratio equal to 5). More importantly, no TCDD was detected in the samples taken from range land having the highest application rate of 2,4,5-T. Of more interest was the fact that four of the above results were from the same southwest Missouri farm. Three others were from two northern Oklahoma farms. The eighth was from a farm in Kansas not treated with 2,4,5-T. Thus, the Dow data showed no relationship between the TCDD content of beef fat and the spraying of 2,4,5-T. At the time, we thought the positive results were caused by poor farm practices. Now, we know faulty hazardous waste disposal practices in Missouri appears to have been the cause.

The analytical chemistry part of Phase I of the Dioxin Implementation Plan was now ended with the following accomplishments:

1. Analytical methodology was developed capable of determining

TCDD in beef fat and liver to a level of 10 ppt.

2. Some criteria were developed for the interpretation of analytical results at very low levels.

3. The analytical methodology was always subject to interferences and was thus incapable of producing unequivocal data.

4. Enough apparent positive results were obtained to suggest the development of more sensitive and specific methods, but not enough to say that TCDD is generally present in beef fat. More importantly, the Plan demonstrated that scientists from academia, government, and industry can work together efficiently even under adversarial situations.

Ralph T. Ross now left EPA. Thomas Holloway and Richard Troast became project managers. The University of Nebraska was now included under contract to EPA.

By the summer of 1977, it became apparent that EPA, in the Jimmy Carter administration, had changed its view of the role of scientists in the Dioxin Implementation Plan. The Analytical Collaborators were no longer consulted on protocol, but were apparently viewed as data generators. The interpretation of analytical data was now done by the EPA. Whether this was politically motivated is not clear. However, at one of the last meetings Maureen Hinkle of the Environmental Defense Fund announced that EDF had decided the scientific route was not the way to go and they were going to use the courts.

As promised in the early stages of the Dioxin Implementation Plan, Dow developed new clean-up procedures. On November 16, 1977, Carolyn Offutt, the new EPA Dioxin Project Manager, Robert Harless, and Aubry Dupuy, visited Midland and reviewed our new methods. At this meeting the emerging sample clean-up methodology was described in detail by Dow scientists who recommended it because it was capable of determining 2,3,7,8-TCDD isomer specifically with no interferences.

On my retirement, Professor Thomas O. Tierman wrote in part:

"I think that in future years, when analytical chemists fully realize the magnitude of advances in the state-of-the-art which were a consequence of the concerns over "Dioxin", they will recog-

nize that many of these achievements resulted directly from
your involvement and scientific direction. What will be less
evident to those who look at this field in retrospect will be the
remarkably successful collaboration between industry, academia
and Government laboratories which occurred in the early
"Dioxin" research programs, and which was due, in no small
part, to your efforts. I regard the period when your laboratory,
along with my own and those of Bob Harless, Pat O'Keefe and
Matt Meselson, and later, Mike Gross, were involved in The
Dioxin Implementation Plan and its successor programs as one
of the high points in my scientific career. I have fond memories
of the numerous meetings of this group at which we argued
about data, exchanged useful information on our laboratory
experiences, and ultimately refined our thinking and capabilities to achieve reliable analyses at the part-per-trillion level.
Surely, we were the first group of analytical chemists to achieve
such capabilities. And even though these meetings were intense working sessions, we occasionally found time for some
memorable social activities."

Indeed, the beautiful working relationships between scientists in
academic, government and industry were models of how such efforts
should be conducted.

DATA, FACTS, AND
HYPOTHESES

"You don't know much", said the Duchess, "and that's a fact",
Alice in Wonderland, Lewis Carroll.

1977

"Just as houses are made of stones, so is science made of
facts, but a pile of stones is not a home and a collection of
facts is not necessarily science", Jules Henri Poincare.

HARVARD UNIVERSITY
DEPARTMENT OF BIOCHEMISTRY AND MOLECULAR BIOLOGY

16 Divinity Avenue
Cambridge, Massachusetts 02138

28 March 1977

Dr. Warren Crummett
Analytical Laboratories
Dow Chemical Company
Midland, Michigan 48640

Dear Warren,

I would like to have your
comments on the enclosed memo and on
other aspects of TCDD monitoring.
Perhaps the simplest thing would be to
telephone after you have had a chance
to read it. My telephone number is
617-495-███.

With best regards,

Matthew Meselson

In any scientific effort in which large collections of data are made, the scientists involved in the study must take special care to reach consensus on several matters, which are interrelated. These include assuring and establishing the reliability and credibility of the data, determining what facts if any can be gleaned from the data, and developing suitable hypotheses consistent with the facts.

For such discussions, I prefer the following definitions:

data: numbers which have been experimentally generated with quality assurance as described in References 2 and 3.

fact: the state of things as conservatively represented by the data.

hypothesis: a provisional supposition from which to draw conclusions that are in accord with the facts. It provides a basis for further investigation.

Scientists involved in the EPA Dioxin Implementation Plan had the difficulty of attempting to assess the reliability of the data and the facts that could be gleaned from these data. Since we were working at the part per trillion concentration level, criteria for data evaluation had not been worked out. So it was very difficult to compare data generated by one laboratory with that of another. With five different laboratories generating data, often by more than one mass spectrometric technique, it was very difficult to know which data were best. To add to the difficulty, different laboratories had somehow acquired different sets of data from EPA, which together with their own, provided the base from which they drew conclusions; that is they attempted to determine the facts.

Because the data set available to Harvard was considerably different from that available to us, their view of "the facts" was considerably different from ours. The data available to Professor Meselson showed a correlation between the concentration in the fat and the rate of application of 2,4,5-T. Data available to us suggested the positive results were a result of poor farm practices. Hoping that these differences could be explained and a consensus reached, I welcomed the letter from Professor Meselson. (In fact, this is one of the most promising invitations I've ever had. If we could agree on the meaning of the data, surely we could persuade the EPA to convene the collaborators and prepare a scientifc paper for publication.) Such discussions are the normal course of science. The situation has been put well by Stephen Jay Gould who said that, "Science has long recognized the tyranny of prior preference, and has constructed safeguards in requirements of uniform procedure and replication of experiments."[1] The DIP researchers had not yet reached a sufficiently uniform procedure and replication of experiments. Because of this situation we were unable to reach an agreement with Harvard before the DIP work was concluded, although we had many useful discussions in which I learned a lot – lessons of importance in future evaluations.

Professor Meselson put the question very well in an interview. "Scientists have an obligation to evaluate evidence according to scientific standards and methods. These are hard-won standards that developed over hundreds of years, in the face of opposition by priesthoods and stoning by villagers. As soon as a scientist begins to tamper with his conclusions for political reasons, the whole system begins to totter."

The EPA DIP Study has never been published and the data from the study are not readily available. The data are complicated due to the different methodology, criteria, and calculations and they are useless for illustrative purposes. On the other hand, data taken by our laboratory was the result of methodology using criteria which was later endorsed by the American Chemical Society.[2,3] So I had complete confidence in its integrity. To illustrate what facts can be drawn from data, we reproduced Dow data from Missouri and Oklahoma cattle, which has been averaged and condensed for easy comprehension.

Dioxin Implementation Plan
Dow Data on Missouri Cattle

Farm	Sample	Age (Months)	245-T Application Rate (lb./A)	Grazing Time (Months)	2378-TCDD(ppt) Fat	Liver
1	1FL	18	3	9	21(10)	ND(5)
	2F	7	3	9	ND(10)	
	3F	36	3	36	61(6)	
	4E	96	3	96	20(8)	
	5E	60	3	60	17(7)	
2	1L	36	3	—	18(7)	ND(10)
	2E	96	3	96	ND(10)	
	3M	36	3	9	9(9)	
3	1L	108	3	108	ND(6)	
	2L	72	3	72	ND(15)	ND(3)
4	1E	108	3	108	ND(6)	
	2F	24	3	24	ND(21)	
5	1F	14	3	14	10(10)	ND(2)
6	1F	13	3	13	21(9)	

E = Emancipated
F = Fat
L = Lean
ND = not detected (limit of detection given in parentheses).

Dioxin Implementation Plan
Dow Data on Oklahoma Cattle

Farm	Sample	Age (Months)	245-T Application Rate (lb./A)	Grazing Time (Months)	2378-TCDD(ppt)	
					Fat	Liver
1	1F	—	2	—	ND(6)	
	2F	—	2	—	ND(10)	
	3F	96	2	96	ND(7)	
2	1L	108	2	10	ND(15)	ND(9)
	2L	108	2	10	ND(16)	ND(3)
	3L	108	2	10	ND(19)	ND(6)
	4L	108	2	10	ND(7)	ND(2)
	5L	108	2	10	ND(6)	ND(4)
3	1F	120	2	10	ND(20)	
	2F	—	2	10	ND(20)	
	3F	108	2	10	ND13)	
	4F	—	2	10	ND(18)	
	5F	—	2	10	ND(3)	
4	1L	—	2	10	ND(5)	
	2L	—	2	10	22(14)	
	3L	—	2	10	30(18)	
5	1F	24	2	10	ND(17)	
	2F	36	2	10	ND(18)	
6	1L	—	2	10	9(5)	
Cont-rols	1F	—	0	—	ND(8)	
	2F	—	0	—	ND(13)	ND(5)
	3F	24	0	—	ND(20)	
	4F	—	0	—	ND(5)	ND(2)
	5L	—	0	—	ND(14)	

E = Emancipated
F = Fat
L = Lean

From these data the following statements of fact can be made:

1. 2,3,7,8-TCDD was not detected in any liver samples
2. 2,3,7,8-TCDD was not detected in fat samples from Texas or Kansas.
3. 2,3,7,8-TCDD was detected in cattle fat from three of six different farms in Oklahoma and four of six different farms in Missouri. Of these, only one sample from one farm in Missouri showed 2,3,7,8-TCDD content sufficiently high to make its determination reasonably sure.
4. There does not appear to be a correlation between 2,3,7,8-TCDD content of beef fat and age, grazing time, nor 2,4,5-T application rate. The condition of the animal is unrelated to the 2,3,7,8-TCDD content.

From these data my hypothesis was that poor farm practices were causing the observed 2,3,7,8-TCDD levels in certain farms in Missouri and probably in Oklahoma. Subsequent events have proven me wrong. Unacceptable waste oil disposal practices were undoubtedly causing the detectable amounts in Missouri. The farms in Oklahoma were also not far from the source of the waste oil.

We were first made aware of the horse arena problem in Missouri in January 1975, when we analyzed a sample of soil submitted by the Center for Disease Control in Atlanta. We reported 32 parts per million 2,3,7,8-TCDD on January 30, 1975. In May we learned more details of the cause of the problem. Contaminated waste oil had been sprayed on the horse arena for dust control.[1]

References:

1.C. D. Carter, R. D. Kimbrough, J. A. Liddle, R. E. Cline, M. M. Zack, Jr., W. F. Barthel, R. E. Koehler, and P. E. Phillips, Science, *118*, 738-740 (1975).
2.W. B. Crummett, F. J. Amore, D. H. Freeman, R. Libby, H. A. Laitiner, W. F. Phillips, M. M. Reddy, and

J. K. Taylor, "Data Acquisition and Data Quality Evaluation in Environmental Chemistry", Anal. Chem., *52*, 2242 (1980).
3.L. H. Keith, W. B. Crummett, J. Deegan, Jr., R. A. Libby, J, K. Taylor, and G. Wentler, "Principles of Environmental Analysis", Anal. Chem., *55*, 2210 (1983).

WHISTLE BLOWING

"There is no education like adversity", Benjamin Disraeli,
British Statesman (1804-1881).

25 August 1975

By telephone, Ralph T. Ross suggested that I bring a Dow attorney to a meeting to discuss dioxin data generated under the EPA Dioxin Implementation Plan because EPA attorneys planned to attend and no one knew why. I was introduced to the group by EPA's chief scientist, Bill Upholt. "This is Dr. Warren Crummett. He represents The Dow Chemical Company." I responded, "Thank you very much, Bill. I would like you to meet my associate. *He* represents The Dow Chemical Company. I represent science!" The EPA attorneys opened their briefcases, unfolded their portfolios, poised their pens and looked at me with keen anticipation. They were sure I would reveal all as I "blew the whistle" on The Dow Chemical Company. I got the message loud and clear that they believed that Dow had many horror stories relating to dioxin and that we were covering them up. It was inconceivable to them that anyone working for industry could be free to conduct sound science and did not have many dark secrets buried deep within their psyche. As for my part, I was stunned to know that it was not evident to them that Dow was an ethical company whose decisions about health and the environment were based on the best data that science could provide. All I wanted to talk about was the behavior of a molecule. What did they really want to talk about? Was Dow viewed as a warlock and I a witch?

Back in Midland I noticed an unusually large number of black-capped chickadees, saucy little much-loved and admired birds at my backyard bird feeder. One of their feature attractions is their call, which has been recorded in most authoritative books on birds as, "chick-a-de-dee" or sometimes shortened to "dee-dee-dee". But the birds in my backyard were extremely happy and excited and were bubbling forth so clearly, there could be no mistake, "tee-cee-dee-dee". For days their song rang through my head. What were they trying to say about TCDD?

In Midland the black-capped chickadees still sing TCDD. If they don't in your backyard and if you don't believe it, come to Midland and listen for yourself.

MYSTERY MAN

"Examine the contents, not the bottle", From the Talmud.

Spring, 1976

A mystery man sat quietly apart from the rest of us; long grayish brown beard flowing over the top of his brand new overalls of denim, long hair creased from a brand new farmers-type straw hat now laying on an adjacent chair, feet encased in clodhoppers planted squarely on the floor and arms clad in the long sleeves of a new blue cotton shirt resting lightly on the arms of his chair. With bright eyes shining straight ahead, he appeared oblivious to the proceedings. At the first coffee break, I approached him and attempted to start a conversation. Furiously, Maureen Hinkle, representative for the Environmental Defense Fund, an environmental activist group, broke off a conversation with Professor Meselson and scolded me, "Don't heckle my scientist!" she said. "He comes here at his own expense to help our cause. Isn't that wonderful?" She explained that the Environmental Defense Fund invites various scientists to give of their time to keep science straight in the Funds projects.

"Why have you never invited me", I asked. "Or have you never had a project requiring knowledge based on scientific data?"

"Someone from industry", she said. "No way!"

"I am interested in your scientist", I said, "because as a lad I wore the same kind of clothes. Seldom did I have a set of new ones, however. The brim of my straw hat, as well as the crown, was frayed, my denim overalls had holes at the knees and buttocks. My shoes gaped in the front where the soles had come loose. My shirt was usually stained under the armpits with sweat. The looser fitting clothes were cool and discouraged chiggers to bore into the skin. One could perspire freely and be aired out so that B.O. did not develop so readily. Nevertheless, I disliked the life and studied hard to escape by becoming a scientist. So I would like to know how someone who has studied as I have was able to live in both worlds."

During the rest of the meeting, which lasted several hours, our mystery man uttered no words. I have not yet discovered who the EDF's "scientist" really is.

MARINE BORERS

"What we see depends mainly on what we look for", Salada tag line.

Summer, 1976

The voice on the telephone sounded urgent. "This is Patrick Q_ _ _ _ of the Port Authority of New York. The damned EPA has cleaned up the Hudson River raising the oxygen content of the water in the harbor from 2 to 6 parts per million. The marine borers are back! They're eating up the wharf. I must stop them! I've searched high and low for an effective poison and I believe I've found one. Yesterday I attended a scientific conference at which the most poisonous substance made by man was described. It is insoluble in water, degrades in light, and does not migrate in the environment. It is called 2,3,7,8-Tetrachloro-dibenzo-p-dioxin. What I want to know from you is where I can buy some and how to handle it without killing myself."

I could not be sure that this voice was talking about a real concern or if it were some kind of joke. Nevertheless, I replied, "I would be happy to tell you how to handle it and you can always find someone to prepare it. But first you must get a use permit from the United States Environmental Protection Agency."

"Whom do I call?"

"The EPA is very interested in this compound – so much so that they have named a very special person as Dioxin Program Coordinator. He is Dr. Ralph T. Ross! His telephone number is (202) 447-_ _ _ _."

"Thank you very much!"

The following day Dr. Ross called me. "You'll never guess the request I've just had," he laughed.

"Did you grant the permit?" I inquired.

"Why you dirty son-of-a-bitch!" he exclaimed.

"Ralph, I only wanted you to know that all things are relative. Whether a chemical is good or bad depends on the urgency of the problem that one has to solve!"

"You taught me that now. You don't ever need to teach it again."

Ralph is a really good man, but he has yet to say if the request were granted.

VISUALIZING SMALL NUMBERS

"Recently, I've had a lot of trouble conceptualizing any number
larger than a million trillion billion",
S. Lams, Cartoon.

1976

In the early 1970's I was asked by lawyers, doctors, journalists, professors, and other scientists to explain the units of "parts per million", "parts per billion", and "parts per trillion" in terms of things seen or experienced in everyday life. I attempted to do this by listening to various statements made by colleagues and peers, carefully checking the calculations, and listing selected ones according to units used in making scientific calculations. The 1976 table is given here.

Trace Concentration Units

Unit	1 ppm	1 ppb	1 ppt
Length	1 inch / 16 miles	1 inch / 16,000 miles	(a six inch leap on a journey to the sun)
Weight	1 ounce / 31 tons	1 pinch salt / 10 tons potato chips	1 pinch salt / 10,000 tons of potato chips
Volume	1 drop vermouth / 80 "fifths" gin	1 drop vermouth / 500 barrels gin	1 drop vermouth in a pool of gin covering the area of a football field- 43' deep
Area	1 square foot / 23 acres	1 square inch / 160 acre farm	1 square foot in the state of Indiana
Action	1 bad stroke / golfer in 3,500 golf tournaments	1 bad stoke / golfer in 3,500,000 golf tournaments	1 bad stoke / golfer in 3,500,000,000 golf tournaments
Quality	1 bad apple / 2,000 barrels	1 bad apple / 2,000,000 barrels	1 bad apple / 2,000,000,000 barrels
Rate	1 dented fender /10 car lifetimes	1 dented fender / 10,000 car lifetimes	1 dented fender / 10,000,000 car lifetimes

Although this table is of little scientific importance, it immediately gained considerable attention in the media. Parts, or all of it, appeared in newspapers, magazines, and trade journals. The impact can be easily seen from the headlines accompanying the table. Some of these are:

"Keeping Millions from Trillions"	Cincinnati Enquirer Bob Brumfield
"PPM-The Mini-Yardstick"	Chemist Analyst June, 1977
"Down to Splitting Frog Hairs"	Phoenix Gazette January 29,1977
"A Finely Split Hair"	Oil and Gas Journal March 21, 1977 Ted Wett
"One Part Per Trillion is a Very Finely-Split Hair"	Chemecology December, 1976
"A Part Per Trillion"	Chem 13 News February, 1978
"A Part Per Trillion—How Big Is It?"	Muskegon Chronicle September 6, 1981 Michael Lewis

Each of the authors added significantly to the understanding of the meaning of these very small quantities. Still, the amount of anything listed as a part per trillion is extremely difficult to visualize. The table with interesting discussion has also appeared in significant books. Among these are:

- M. R. Wessel, "Responsibility, In Science", Chapter 6, "Science and Conscience", Columbia University Press, NY, pp 124-129, (1980)
- C. E. Kupchella and M. C. Hyland, "Environmental Science. Living Within the System of Nature", Allyn and Bacon, Inc., Second Edition, p 287, (1989).

- W. D. Rowe, et. al., "Evaluation Methods for Environmental Standards", CRC Press Inc., Baca Raton, FL.

Now we are being asked to stretch our minds further still as government agencies ask for measurement at parts per quadrillion or even part per quintillion. If we attempt to include these quantities in our table the examples become incomprehensible for many of us. The following table is an attempt to extend the first table. Perhaps you can fill in the blanks with meaningful examples.

Unit	1 part per quadrillion	1 part per quintillion
Length	1 inch in 100 round trips to the sun	1 foot in a million round trips to the sun
Volume	1 drop of vermouth in a pool of gin covering the area of 1000 football fields 43' deep	1 drop of vermouth in a pool of gin 43' deep covering the area of one million football fields
Area	0.6 square foot in all of the earth's land area	0.08 square inch in all of the earth's land area

It is important to realize that scientists use the following units to describe parts per million, parts per billion, etc.:

Part Per Million =	1 microgram per gram
Part Per Billion =	1 nanogram per gram
Part Per Trillion =	1 picogram per gram
Part Per Quadrillion =	1 femtogram per gram

To put this in perspective, a femtogram is 0.000000000000001 gram, while a microgram is 0.000001 gram. Neither of these quantities is large enough to be seen by the naked eye.

ENCOUNTER WITH A
PROFESSOR

"We live on Earth and not in heaven, and Earth is a dangerous place", Magnus Pyke.

23 June 1978

Walking into a New York Academy of Sciences President's Reception for the first time is like entering a deep freeze. Small groups of "in" scientists stand about giving newcomers icy stares. The one on June 23, 1978, was like that. Surprisingly, a lady approached me, introduced herself, and asked me where I was from. When I revealed I was from Michigan, the dialogue became intense.

"I would not set my big toe anywhere in the state of Michigan", she said.

"Why not? It's a great place!" I responded.

"Because Michigan has PBBs and that's reason enough."

"You shouldn't condemn the entire state for that. The contamination at levels of health concern is not wide spread."

"Well, then why did Dr. Selikoff go to Michigan? You can't tell me that he would have gone if the situation weren't serious."

"I'm sorry, but I don't agree with you. Dr. Selikoff, like other scientists, will go wherever there is an opportunity to apply the scientific method to a subject of interest. To put things in perspective, let's talk about New York. When I landed at the airport I noticed strange odors and as the taxicab brought me into the city these odors changed, many becoming more intense. There is no doubt whatsoever that you and I are standing here in a sea of carcinogens, and . . ."

"Oh, my God!" Her knees appeared to waver and she mumbled,

"See you later."

Twenty minutes later she was back. "There are some people I want you to meet", she said. We walked across the ballroom and I was introduced to two toxicologists, one from a major California university, the other England. Other scientists were also in the group. "Tell them what you told me!" Her voice expressed the triumph that she appeared to feel.

"We are standing here in a sea of carcinogens!" I said a bit unsure of myself.

"You bet we are!" they laughed.

"Oh, my God!" she said. "You mean it, don't you?" She sounded incredulous.

"And relatively few of them are manmade," someone interjected.

She calmed down. "I have learned more in the last twenty minutes than from years of reading the New York Times, the Wall Street Journal, Time, Newsweek, etc." Now on casual terms, we chatted about other things. She was a professor of mathematics at one of New York City's most prestigious colleges.

Enough information has now appeared in the popular press that the professor would be properly informed. Edith Efron[1] has summarized the available information on naturally occurring carcinogens and Bruce Ames[2,3] has put their role in nature in perspective. Humans are exposed to these natural materials through radiation, the earth's crust, viruses, bacteria, fungi, plants, food, air, water, fire, and man.

Air, which we were inhaling and which constituted the "sea of carcinogens" in which we were standing, contains various trace amounts of radionuclides, soil, sand, dust, arsenic, cadmium, copper, lead, iron, manganese, mercury, selenium, zinc, chromium, cobalt, nickel, a-pinene, isoprene, benzo(a)pyrene, nitric oxide, nitrous oxide, sulfur dioxide, and ozone. All of these are produced naturally and all have been said to be carcinogens. In addition, it contained the special odors of New York City, including the synthetic ones associated with hotel construction and decoration. But, more importantly, it contained the aromas from the preparation of a wide variety of foods.

Still, my new friend would not have been fearful if she had been reading Science magazine. In this journal an imminent scientist, H. E. Stokinger[4], Chief of the Laboratory of Toxicology and Pathology in the National Institute for Occupational Safety and Health put this matter in true perspective. "Physiologically, man needs continual toxicological nudging to maintain the homeostatic mechanisms that keep him physically and mentally alert." "Man has never, before he was a man or ever after, survived in an unpolluted void."

References

1. Edith Efron, "The Apocalyptics", Chapter 4, "Benevolent Nature", Simon and Schuster, New York (1984), pp 123-183.
2. Bruce N. Ames, "Cancer Scares Over Trivia", Los Angeles Times, May 15, 1986.
3. Bruce N. Ames, "Peanut Butter, Parsley, Pepper and Other Carcinogens", The Wall Street Journal, March 17, 1984.
4. H. E. Stokinger, "Sanity in Research and Evaluation of Environmental Health", Science, 174, pp 662-665 (1971).

THE COMMITTEE THAT *WUZ*

*"Advice is seldom welcome; and those who want it most always
like it least",*
Phillip Dormer Stanhope, 1694-1773.

1976-1978

On March 26, 1976, Russell E. Train, Administrator, U. S. Environmental Protection Agency wrote "delighted to invite you to serve on the Environmental Advisory Committee of the U. S. Environmental Protection Agency for a term beginning immediately and ending June 30, 1978, subject to prescribed appointment procedures". I was very pleased and accepted immediately after consultation with Dow management, who readily concurred. I was elated when I learned the composition of the committee:

Chairman: Frederick D. Rossini, Professor of Chemistry, Rice
  University
Executive Secretary: Alfonse F. Forziati, Staff Scientist, EPA
  Science Advisory Board
Members: Lenore S. Clesceri, Department of Biology, Rensselaer
  Polytechnic Institute; John O. Corliss, Department of
  Zoology, University of Maryland; Ursula Cowgill, Department of Biology, University of Pittsburgh; Bryce L.
  Crawford, Department of Chemistry, University of Minnesota; Edward F. Ferrand, New York City Department for Air
  Resources; Virgil H. Freed, Department of Agricultural
  Chemistry, Oregon State University; Henry Freiser, Department of Chemistry, University of Arizona; Choo-Seng Giam,
  Department of Chemistry, Texas A&M University; Joel H.

153

Hougen, Chemical Engineering, University of Texas; Edwin H. Lennette, State Department of Public Health, Berkley, California; James H. Pitts, Jr., Statewide Air Pollution Research Center, University of California; (Riverside) Lockhart B. Rogers, Department of Chemistry, University of Georgia; William C. Taylor, Civil Engineering, Howard University; Goeffrey F. Watson, Department of Statistics, Princeton University; George Zissis, Chief Scientist, Environmental Research Institute of Michigan

All of these members had distinguished themselves in their respective fields. As the only member from industry I felt inadequate to appropriately benefit from their vast knowledge and wisdom. However, I soon learned that they were as eager to learn from my experience as I from theirs.

Our function was to provide reviews of the overall environmental measurement activities of the agency; the appropriateness and effectiveness of the research effort; priorities, feasibility—everything that can be considered in the operation of a laboratory whose function is measurement .

To accomplish our mission sub groups of the committee visited many of the agency's laboratories. There we listened to the goals, accomplishments, plans, and vision of the various groups and did walk-through audits of the laboratories. I participated in nine such visits to laboratories located in; Athens, Georgia; Cincinnati, Ohio; Corvallis, Oregon; Duluth, Minnesota; Edison, New Jersey; Narragansett, Rhode Island; Gulf Breeze, Florida; Research Triangle Park, North Carolina; and Seattle, Washington. These visits were highly educational. Generally we were treated to detailed discourses on the function and operation of the laboratory based on elaborate rules and procedures. So it was generally easy for us to make suggestions about cutting "red tape" to gain efficiency of operation, and to enhance effective communications.

However, the few such trips that I did not take proved to be the most unusual and revealed the character of the committee and, to some extent, that of the agency. The first trip, taken by others to Las

Vegas, resulted in a report, which lauded the capabilities and performance of the laboratory. Some who had made the visit told me individually and privately that the report failed to describe the true situation. So, when the report was submitted for approval by the entire
committee, I commented that I was sorry that I had not gone as I had
never visited a laboratory so well constituted and operated. At that a
vigorous debate ensued which lasted the rest of the afternoon, through
dinner, and on into the night. The few, who had apparently written
the report, held the view that, in order to be sure the laboratory
could continue to be funded, we had to paint a picture of extraordinary competence. Others held that we had to report the truth as we
saw it. The matter was finally resolved by comments of William Taylor who said that the report, was like one written by a salesperson,
who had not made a call. So the report was rewritten, and subsequent reports included harsh criticism as well as high praise.

The die-hards, who did not trust the administration with the
truth, seemed to lose interest in the committee. Others commented
to me that we had an interesting discussion but it was a terrible waste
of time. My perception, which I shared with them, was that we had
come together as a group and could now deal openly and honestly
with each other. Apparently such open discussions rarely occurred in
academic circles. But at Dow they were daily events.

I did not make the visit to the Chicago Regional Laboratory on
May 12-18, 1978, with the subgroup. I declined to go because various representatives of that laboratory had made unfounded disparag-
ing remarks about The Dow Chemical Company and our dioxin work
and I thought some would say I had a "conflict of interest". However,
the subcommittee had a very negative experience and wrote a scathing report. This report was distressing to both the management of
the laboratory and to the head of EPA's Science Advisory Board, Dr.
Richard Dowd. He called and persuaded me to audit the Chicago
laboratory, accompanied only by him. On June 2, we found that the
laboratory had made adjustments and now flagrant inadequate practices could no longer be found. So I wrote a positive report. However
it seemed ironic that I was the one to return credibility to a laboratory that was trying to unjustly persecute me and my people.

After having visited a sufficient number of EPA laboratories, the committee felt the need to visit a reference laboratory and decided to visit the Dow laboratories. There they toured and audited Toxicological Research, Environmental Sciences Research, Agricultural Products Research, Organic Analysis, Inorganic Analysis, and Environmental Sciences groups. In Midland they had dinner at the Shanghai Peddler and were served a specially prepared meal by Mrs. Carl Chow, later of Bamboo Garden fame. Inspired in the dinner's afterglow, John Corliss wrote the following poem which he submitted as, "my two-year report!".

Ode (owed) to The Committee That Wuz
(A Lament, delivered spontaneously on the Occasion of our Last
Supper,
Midland, USA, 16 May 1978).

Here's to our Committee, forlorn and quite lost.
Nobody loves us, though demonstrably low is our cost.
We've traveled far and we've traveled wide
And very few if any seem to be on our side.
Tis' well-known that we're anti-social and rude,
That may not be bad, but it's not 'xactly good.
So we're self-centered and a bit anthropomorphic,
Sleepy at times, call us Super-soporific.
We write reports, but nobody'll listen
'Till we rant and roar and keep on insistin'.
The impact of our wisdom (collective) may never be known,
But we've tilled the field and our seed has been sown.
And if EPA is ne'er agin quite the same,
We the Committee proudly'll 'sume all the blame!

John O. Corliss

Indeed EPA was never the same, but the agency did not take the advice given in our report. By the time the report was issued the politics of the country had changed. Jimmy Carter was now president and Douglas M. Costle was EPA Administrator. There is no evidence that either read the report. The Administrator did write a belated letter of appreciation October 2, 1978, but it showed no indication that the report had been considered.

Among the topics discussed in the final report, the following are especially noteworthy:

The Problem of Measurement

"The problem with measurement is basic to the needs of EPA in its regulatory responsibilities and in educating its constituencies to a better understanding of the handling of environmental problems. The science and technology of measurement are fundamental to achieving and maintaining the desired quality of the environment. It is important to maintain the integrity of the process of measurement throughout, from the inception of a method to its scrutiny in the courtroom. The basic elements of the measuring process include: the collection of valid samples; the development and use of standard methods of test and their associated standard reference materials; and maintenance of a quality assurance program through intercomparison of results by different laboratories via round robin measurements. There is need to emphasize that every measurement of whatever kind has associated with it a range of uncertainty. This range of uncertainty is an integral part of the measurement and should be so reported. There should be uniformity in the testing and reporting of the same pollutant, whether it be in detecting, monitoring, control, assaying, or whatever, in different media. There should be an Agency-Wide Center for standard reference materials for all media, whether air, water, or soil. As previously emphasized, a central authority of the Agency is needed to assure quality control of measurements through appropriate use of standard methods of test, standard reference materials, and so forth. As emphasized again and again, the problem of measurement is basic to the entire EPA enterprise: unless we can measure it, we know nothing about it."

The Problem of the Unattainable Absolute Zero Pollution

"The problem here is, in fact, quite simple: There is no such thing as absolute zero pollution on this earth; there is no substance whatsoever that has absolutely zero impurity in it, nor can man ever make it so. Unfortunately, some scientists and engineers, as well as members of the U. S. Congress, are unaware of this fact, which is a result of the immutable laws of physics and chemistry and the consequence of the containment of substances. What must be appreciated are the needs for the determination of the level of the given pollutant in the given environment that can be tolerated without significant adverse effect on man or the environment. It is very unfortunate that there exists in the Delaney Clause, in one of the laws enacted by the U. S. Congress, the implication that there can exist an absolute zero level of contamination. The unattainability of the absolute zero pollution is similar to the unattainability of the absolute zero of temperature. Man can approach the absolute zero of temperature ever so closely, *at increasingly greater and greater cost*, but can never actually reach it."

The Problem with the Unattainable Zero Risk

"Here again, the problem is simple: There is no human activity of any kind whatsoever that has absolute zero risk associated with it. The U. S. Congress and U. S. Government regulators need to keep the foregoing statement clearly in mind along with the following: The human enterprise is a complex system, consisting of an almost infinite number of entities which have developed unique individual modes of existence based on local and neighboring conditions and constraints. No one entity is independent of all others. Each is related in some one or more facets of its existence to neighboring entities and conditions. Alteration of one condition in one entity to improve its condition may alter the condition of many other entities, for better or for worse. Removal of one risk from one entity may freely expose that entity to another greater risk. What it all boils down to is that, in a complex system, one does not alter one condition without carefully considering the total effect of that alteration, for good or for bad. Every action has associated with it a risk. The problem is simply to try to assess the relative risks of several options, considering always the complexity of the human enterprise."

Shortly after the final report was written, I received a mysterious telephone call. The caller identified herself to my secretary as a young Ph.D. scientist with the EPA in Washington, DC. She left her telephone number but requested that I not return her call as my name was "mud" in the agency and she would jeopardize her career if it became known that she was in communication with me. Later she succeeded in reaching me by telephone. She had heard that the Environmental Measurements Advisory Committee had written about how professional chemists should be managed. She had been unsuccessful in obtaining a copy of the report through the Agency. She was concerned that women in the Agency were not encouraged to attend scientific conferences. So I sent her a copy of the 1977 report which had discussions entitled, "Philosophy and Organization for Research and Development", and "Management at Headquarters and at the Laboratories".

On September 20, 1977, at the invitation of Dr. Alfonse Forziati, later Staff Director of Biological and Climatic Effects Research Program (BACER), I sat as an evaluator and critic near the front of the meeting in Washington, DC. While there, I was approached by a beautiful young lady who identified herself as the EPA scientist who had contacted me previously about the EMAC report. She quickly thanked me and hurriedly departed. She still thought that being seen in contact with me would be injurious to her professional advancement. That opinion did not change after I assured her that I had been thoroughly investigated and had an official EPA document stating that there was no conflict of interest between my Dow job and consulting for the Agency. She still thought the Agency viewed me as the enemy. I've had no further contact with her but other contacts suggest that her efforts to upgrade the role of scientist in the agency were not successful. I still believe that each Agency administrator should read our report with its basic message for management.

The most curious thing about this committee was that it was given no publicity. No reporters were ever present and neither were environmentalists. I was never approached about the committee or its business.

Viewed as a consultant and confidant by the EPA research scientists but an enemy by some administrators and bureaucrats in the Agency, I could not have reached this position by action of the dioxin molecule alone, but some powerful sinister force must have been operating. Thinking about it caused the following rhyme to gush forth:

Requiem for the EPA
Environmental Measurements Advisory Committee
(1976-78)

What a pity?
The Committee
Labored oe'r the nitty-gritty,
Without a battery of laws,
To give administrators cause,
For reflective pause!

Gently teaching,
Never preaching,
Using words of hope beseeching;
We put reliance,
In the Science,
But not just to meet compliance!

In Midland City,
The Committee,
Focused on the itty-bitty;
Zero risk is just a fad!
Zero pollution cannot be had!
Data uncertainty? – at least a tad!

Parts per million,
Billion, trillion,
Slip sliding toward a part per zillion;
Environmentalists are sure,
'Tis the only way to make things pure,
So that weak genes can endure!

Fully smitten,
We found it fittin'
That the report be wisely written;
But to this day, the EPA
Can't say where 'twas filed away.

ON THE SPOT

*"I was gratified to be able to answer promptly, and I did. I said I
don't know",
Mark Twain, Adventures of Tom Sawyer, 1976.*

Winter 1976

I was taken aback by the enthusiastic, warm greetings of Dow'
top research managers. Soon I knew why. Each had a "secret" question, which appeared to be on everyone's mind. "Has dioxin really
been found in Tittabawassee River fish?"

I answered openly and honestly. "I don't know! If the Michigan
Division has such information they will inform you at the appropriate time." I had heard the same rumors they had, but, even though
the study was being conducted in my laboratory, I had not seen the
data. So, I would not add to the rumor until I was satisfied that the
data was valid.

At least five different top company research managers, individually and secretively, asked me this question. Each expressed a special
need to know. My answer remained the same. I had seen no data.
Obviously some did not believe me, but it was the truth.

They were all disappointed. Some insisted and didn't believe me.
I now knew why I had been invited to this special gathering at the
home of Dow's Director of Research, Dr. Julius Johnson. I worked for
the Michigan Division but had many contacts in the Corporate function. Each of them had special friends in high places outside Dow
who had "a need to know" before the news media.

A few days later the analyses were complete. Dow chemists had
found parts per million levels of polychlorinated biphenyls (PCBs)
and polybrominated biphenyls (PBBs), and parts per trillion levels of

polychlorinated dibenzo-p-dioxins in fish from the Tittabawassee River. This information was discussed with the Michigan Department of Natural Resources (MDNR). They asked for a few days to consider the data and formulate a policy, after which they would hold a joint press conference with Dow. Unfortunately the very next morning the news was in many Michigan newspapers with scary headlines. The articles featured the discovery of dioxins and down played the PCBs and PBBs although at the amount found, they should have been of more concern.

If I could help it, this would not happen again. I felt I had done the right thing, but Division management had been too trusting.

IDEAL ANALYTICAL SYSTEM

*"We have developed the science to such a degree that we can
hardly be sure that some of the elements we are finding might not
have been there all along", Rev. Carl E. Price, a sermon entitled,
"Traces of Jesus".*

An ideal analytical system would be capable of determining each
chemical compound specifically (without interference from other isomers, congeners, or other related compounds) at limits of detection
ten times lower than the lowest concentration level of interest. The
2,3,7,8-TCDD isomer can be determined specifically. However, a
concentration level below which there is no concern has not been
established for any type of environmental sample except fish. If the
FDA limit of 25 parts per trillion of 2,3,7,8-TCDD in fish is accepted for regulatory purposes, then the analytical methodology should
be isomer-specific with a detection limit of 2 parts per trillion. The
same logic would apply to any other individual molecule one wished
to determine.

In reality only a few laboratories are capable of determining the
2,3,7,8-TCDD isomer specifically. However, many laboratories are
capable of measuring signals at a limit of detection below one part
per trillion. If interferences from other isomers and similar compounds
are not removed, the results are often false.

Often, different laboratories will report data, which are in sharp
disagreement. The worst case is that in which one laboratory reports
a positive result while the other reports that dioxin was "not detected".
In such cases, the supplier of the sample is confused – the only recourse is to employ a third laboratory and perhaps a consultant. Even
so, the debate can go on for years.

MEANING OF ISOMER SPECIFIC

"Dioxins are only chemicals. There is nothing supernatural or
monstrous about them",
Donald G. Crosby, 1983.

Several families of chlorinated organic compounds have been found
in the environment at very low concentration levels—generally below parts per billion. These include the biphenyls, terphenyls, naphthalenes, anthracenes, phenanthrenes, and fluorenones, as well as the
dibenzo-dioxins and dibenzofurans. Each of these consist of groups
of isomers depending on the number of chlorine atoms in the molecule. An isomer is a molecule whose atoms are identical in kind and
number with those of at least one other molecule. The possible number of isomers for a particular number of chlorine atoms is shown in
the table below.

So we have seen that there are two chlorodibenzo-p-dioxin isomers containing 1 atom of chlorine per molecule, 10 containing 2
atoms, 14 containing 3 atoms, 22 containing 4 atoms and so on
through the table for all the various families of compounds. There are
a total of 75 different chlorinated dibenzodioxin molecules. Each of
these is called a congener. Thus, there are 75 congeners of the chlorinated dioxins. Similarly, there are 75 naphthalenes, 135
dibenzofurans, 135 fluorenones, 209 biphenyls, 813 anthracene/
phenanthrenes, and 8557 terphenyls.

Since most of the chlorinated dioxins have been found in some
fly ash, dust, and soil samples at concentration levels greater than a
part per trillion, congeners of the other families are likely to be there
as well. This means that at least 10,000 different chlorinated organic

molecules are present in materials such as fly ash at levels of 1 part per trillion or higher. To determine 2378-TCDD without including any of its isomers or other chlorinated molecules requires that it be separated from all the others. Such an analysis is truly "isomer-specific". Unfortunately, the term "isomer-specific" is used inappropriately by some laboratories to mean that they have looked at a signal where 2378-TCDD usually produces one, as opposed to looking at the signals of all the 22 TCDD isomers. Such claims are often misleading and make the comparison of data taken by different laboratories very difficult.

We know that the molecule 2378-TCDD exists because we have measured it by several special highly reliable instruments. These include single crystal x-ray diffraction, infrared spectrophotometry, nuclear magnetic resonance, mass spectrometry, and elemental analysis. All the measurements fit the structure.

SOME CHLORINATED ORGANIC MOLECULES
Number of Isomers

# of Chlorine Atoms	Dibenzo Dioxins	Naphthalenes	Dibenzofurancs	Fluorenones	Biphenyls	Phenanthrenes	Terphenyls
1	2	2	4	4	3	8	15
2	10	10	16	16	12	40	77
3	14	14	28	28	24	92	234
4	22	22	38	38	42	170	576
5	14	14	28	28	46	192	1035
6	10	10	16	16	42	170	1504
7	2	2	4	4	24	92	1672
8	1	1	1	1	12	40	1504
9					3	8	1035
10					1	2	576
11							234
12							77
13							15
14							3
Total Congeners	75	75	135	135	209	814	8557
						TOTAL	10000

THE METHOD SUPREME

"Analytical chemistry is a process in which you can do a hundred things right and only one thing wrong and still come out with the wrong answer",
Ralph E. Oesper, Emeritus Professor.

In order to appreciate the elegance and rigor of analytical methods, developed and fine tuned by the *Wizards*, and to understand why analysis is so time-consuming and difficult, the entire procedure must be examined in some detail.

By 1990, the recommended method would produce data specific for 2,3,7,8-TCDD with typical recoveries of 75 percent and approximate limits of detection as follows:

Sample	Sample Size, grams	Limit of Detection part per trillion
Water	4000	0.001
Particulate Matter	5	1.
Fish and Shrimp	20	0.2
Human adipose tissue	5	0.5
Cod liver oil	5	0.8
Animal feed	20	1.

Such results, as yet unmatched by any other analysis, were achieved through meticulous commitment to details and developing a thorough understanding of what each step in the process does and what it leaves undone. If any step in the process is done incorrectly, the integrity of the entire analysis is put in jeopardy and the data cannot be reasonably defended. This is why different laboratories seldom produce comparable data.

The analytical method outlined below was developed through the creativity and innovation of the *Wizards*, Terry J. Nestrick and Lester L. Lamparski, based on the early work of Rudolph Stehl, Richard A. Hummel and Lewis A. Shadoff.

First, the samples must be taken for the explicit purpose of analysis for dioxins. The samples must be protected from contamination by dioxins and other interferences normally present in the environment. Therefore, the sampling devices and sample containers must be free of all trace organic materials. Rubber or plastic seals are unacceptable. The exact location of the sampling must be recorded as well as the time. Photographs are useful to record the event.

Second, the custody of the samples must be documented. (It is best for the analytical chemists to do their own sampling).

Third, sample preparation consists of removing the oil-soluble material suspected of containing the dioxins, from the bulk of the sample. This is accomplished through extraction and filtration. Great care must be taken to avoid contamination of the oil extract. The laboratory must be scrupulously clean. All particulate matter must be excluded from the laboratory by filters, positive air pressure, and shoe-cleaning devices at all entrances. Isotopically-labeled internal standards are used to measure analyte recovery.

Fourth, the extract is cleaned up through a series of three low-pressure liquid chromatography columns. The first column consists of layers of activated silica, sodium hydroxide on silica, activated silica, sulfuric acid on silica, and activated silica. The second column consists of silver nitrate/silica and the third is basic alumina. Collectively these remove acidic species, polar compounds, basic species, terpenes, many polynuclear compounds, many compounds of low solubility, and easily oxidizable compounds.

At this point the effluent represents about 1 part per million of the original sample. To determine dioxins at the part per trillion level, 1 part must be separated from another million parts. The separations have already required the use of about 100 pieces of glassware per sample. Each piece was cleaned twice with a thoroughness which would not permit contamination by even 1 picogram of the dioxin to be determined.

Fifth, various analytes are separated on a reversed phase, high performance liquid chromatograph (RP-HPLC). The fraction containing 2,3,7,8-TCDD is carefully collected.

Sixth, further separation is achieved by use of normal phase, high performance liquid chromatography (NP-HPLC).

Seventh, an aliquot of the effluent is injected into a gas chromatograph mass spectrometer. (The performance of the mass spectrometer has been optimized over the mass range of interest. It has been calibrated daily for each analyte of interest, and is recalibrated after every four sample analyses.)

Eighth, positive-identification of the analyte is made if the following criteria are met:

1. The RP-HPLC elution zone corresponds to that of the standard
2. The NP-HPLC elution zone corresponds to that of the standard
3. The correct retention time is found from the gas chromatograph
4. The parent ion-halogen isotope ratio is correct
5. The signal to noise ratio on the mass spectogram is equal to or greater than 3
6. A fragment ion from the loss of COCl is observed
7. Recoveries of the internal standard fall between 50 and 125%.

When all the qualitative criteria are met, quantitative calculations can be made and appropriate tables of data can be constructed. In the 1970s, two exceptional analytical chemists working as a team could analyze about 3 samples per week at a cost of about $1,000. per sample.

No other molecule or family of molecules has been subjected to such elegant, complex and rigorous analytical measurement. In both qualitative and quantitative performance, these methods stand apart from all others.

In spite of the magnificence of this methodology, the limit of detection and reproducibility can only be guaranteed by the Dow scientists who did the development. When the data are compared with those of another laboratory the Horwitz Equation for precision is likely to pertain:

$$V(\%) = 2^{(1-0.5 \log C)}$$

Where V is the reproducibility coefficient of variation and C is the concentration expressed in powers of 10 (for example, 1 ppt = 10^{-12} and log C is -12.) So at a level of 1 part per trillion the coefficient of variation would be expected to be 2^7 or 128%. Thus at a level of 10 ppt the data might range from "non-detected" to 22 ppt. This makes regulation extremely difficult.

In 1982 I was invited to present an earlier version of this methodology at an international symposium entitled, "Pesticide Chemistry – Human Welfare and the Environment", sponsored by the International Union of Pure and Applied Chemistry in Kyoto, Japan. The presentation emphasized the importance of understanding the function of each step in the process and applying the appropriate controls to each to assure top quality of the data. The poster was well received by the conference attendees, especially the Japanese.

On our return to the United States, my wife and I spent a night in Los Angeles. At the airport the following morning we noticed a man handing out pages which he tore from a book. Many passengers turned their heads and walked quickly by. Feeling some compassion I accepted an offering and stuffed it in my pocket. On boarding the plane I pulled this paper from my pocket and was amazed to see a Bible verse which struck me with great force, "A false balance is an abomination unto the Lord but a just weight is his delight". It was the essential message I had taken to Japan. In spite of my scientific mentality this happening made me feel very good. A few years later I told this story in a talk at the National Bureau of Standards in Gaithersburg, Maryland. I asked if anyone would calculate the probability of my getting this special verse. Both publicly and privately scientists said they could not. I asked if anyone could give me a scien-

172

WARREN B. CRUMMETT

tific explanation. No one could. I then pointed out that interpretation by scholars in different disciplines would be vastly different. Everything from miracles to mere chance would be concluded.

Could my contact with this Bible verse be linked to dioxin? Using the logic of some risk assessors and the media, I have to say, "Yes".

References

1. W. B. Crummett, T. J. Nestrick, and L. L. Lamparski, "Analytical Methodology for the Determination of PCDD's in Environmental Samples: An Overview and Critique", "Dioxins in the Environment", M. A. Kamrin and P. W. Rodgers, Eds., Hemisphere Publishing Corporation, N. Y., 1985.
2. L. L. Lamparski, T. J. Nestrick, and W. B. Crummett, "Determination of Specific Halogenated Dibenzo-p-dioxin and Dibinezofuran Isomers in Environmental and Biological Matrices by Gas Chromatography–Mass Spectroscopy", "Environmental Carcinogens Methods of Analysis and Exposure Measurement", Vol. 11, Polychlorinated Dioxins and Dibenzofurans", C Rappe, H. R. Buser, B. Dodet, and I. K. O'Neill, Eds., IARC Scientific Publication No. 108, Lyon, 1991.
3. William Horwitz, "The Evaluation of Analytical Methods Used for Regulation of Food and Drugs, U. S. Food and Drug Administration", Washington, D.C., Private Communication.
4. W. L. Horwitz, L. R. Kamps, and K. W. Boyer, "Quality Assurance in the Analysis of Foods for Trace Constituents", J. Assoc. Off. Anal. Chem.

TRACE CHEMISTRIES OF FIRE

Discovery of An Hypothesis

"In framing hypotheses we must see that they agree with the facts; in other respects, they may be as inconceivable (not self-contradictory) as any fairy tale", M. M. P. Muir.

1978

"Here it is! See if you can make some sense of it!" The representative from the Dow Dioxin Task Force instructed me as he delivered a large file of data. I was aware that a Dow team of about 30 people had been working for several months to determine the sources of 2,3,7,8-TCDD found in the Tittabawassee River fish at low parts per trillion levels. The trace levels of dioxin in the fish had been discovered by Dow scientists in 1976. Other contaminants at much higher concentration levels were also found, including polychlorinated (PCBs) and polybrominated (PBBs) diphenyls. Dow had promised the Michigan Department of Natural Resources that we would identify the source of the TCDD in the fish.

A major investigative effort had been undertaken which examined water in the waste ponds and the river, sediment from the ponds and river, and effluent from plants where chlorinated phenols and their derivatives were manufactured. Although the analytical methodology was the most sensitive of any known at the time, we were unable to show a relationship between the plants' operations and TCDD in fish in the river. So we developed new analytical methodology which was much more sensitive and specific for individual compounds. In the course of this development, we found interferences in

173

our blank samples. This led Dow scientists to examine the dust condensed in the air filters of the laboratory. The dust was found to contain many of the chlorinated dioxins. This suggested that the dioxins at the Midland site were airborne. This was a significant finding and caused the investigation to focus on dust, fly ash, and soil.

Everywhere the Dow scientific investigators looked in Midland, or elsewhere, the dust, fly ash, and soil from inside or close to combustion facilities were found to carry many of the chlorinated dioxins at part per billion levels. The data documenting this were in the package handed to me.

Afterwards, I was named a member of the Dioxin Task Force, chaired and directed by Robert Bumb. I was soon to learn that this group was unique. Everything discussed was considered confidential and could not be discussed with anyone outside the group, not even other task force members.

The task force included two organic research chemists, three analytical research chemists, two attorneys, two journalists, two incinerator experts, an industrial hygienist, a chemical engineer, a businessman, a secretary, and the director of research. All decisions required unanimous agreement. Since this was difficult to achieve, the discussions were long and vigorous.

The data given to me had been taken by three different analytical methods, each different with respect to sensitivity and specificity. Moreover, the sampling had not been done by statistical design. The investigation had proceeded in a manner similar to that of a detective story. Data were taken on a few samples. The results were clues which suggested further sampling and analysis. This process was then repeated. This mode of sampling and analysis made it difficult to detect a common thread on which to base a scientific hypothesis. So I did a lot of thinking and very little writing. The data were convincing in one respect – it showed that TCDD was not being leaked from the production of 2,4,5-trichlorophenol or its derivatives in detectable amounts and did not appear to account for that found in the Tittabawassee River fish. Thus, the hypothesis that this was the source of the dioxins in fish put forth by the Michigan Department of Natural Resources was not adequate—at least it was not the whole story.

On the other hand, dioxins were found on the particulate matter being emitted from the Dow incinerator and the coal-fired power plant. The results obtained from dust and soil from East Lansing, Chicago, and Detroit were all positive with respect to various chlorinated dioxins, dibenzofurans, biphenyls, terphenyls, etc. Thus, it appeared that the dioxins were widespread. Still, I was not yet ready to say that dioxins were always produced from fire.

My problem, of course, was that I was totally concerned with mechanism. I wanted to know the names of all the building blocks needed to form dioxins. It was certainly known that chlorinated phenols, when heated, served as building blocks to produce dioxins[2], especially in the presence of a basic material. And it was reasonable to assume that all chlorinated phenols were produced by industry. Certainly, experiments by Professor Dr. Christoffer Rappe[1] of the University of Umeá, Sweden and coworkers demonstrated that dioxins could be produced from the thermal condensation (pyrolysis) of phenols.

The discovery by the Dow task group that dioxins were present in cigarette smoke and automobile mufflers motivated me to rethink what was happening in a fire. It was well known that salt (sodium chloride) and other metal chlorides [3], when fused at high temperatures would decompose to release chlorine in dry air or hydrogen chloride in moist air. It was also well known that small organic molecules such as methane or ethylene could, when heated, form very low yields of aromatic compounds[4], which could condense in fires to form polynuclear aromatic hydrocarbons. Recently, the process is sometimes called pyrosynthesis.

In the latter case, the yields were very low – generally parts per million or less. Suddenly all was very apparent to me. I had been searching for mechanistic reactions, which require at least ten percent yield. What we were actually dealing with were reactions which proceeded at the parts per million, part per billion concentration levels. Mechanisms for reactions occurring at such low concentration levels had not yet been studied very much. So other mechanisms in addition to the well-known chlorinated phenol condensation would likely play a significant role.

It was well known that combustion processes are seldom more than 99.9 percent efficient in converting the carbon content of fuel to carbon dioxide,[5,6,7] The remaining 0.1 percent of the carbon in the fuel is converted to numerous trace organic compounds including chlorinated hydrocarbons, only a few of which had been identified.[9] Only a tiny fraction of these compounds need be converted to dioxins to account for the miniscule quantities observed in combustion products.

It was also known that the natural chlorine content of fuels ranges from 1 to 5000 parts per million.[8] Whether this chlorine is present as a chloride salt such as sodium chloride (common table salt) or as a part of an organic compound was not clear, nor is it now. Nevertheless, it was well known that hydrogen chloride is produced when these fuels are burned, with yields that are nearly quantitative.[8] Furthermore, it had been demonstrated that chlorine gas and hydrogen chloride are produced when sodium chloride or calcium chloride are fused in a furnace at high temperatures.

The realization that the fly ash or particulate matter in the combustion process could possibly serve as a catalyst for many trace reactions or as a part of the reaction mechanism itself occurred when data were found that proved the presence of nickel, vanadium, iron, and manganese in these particles.[9]

Further thinking along these lines led to a report in which these considerations were pointed out, together with a graphic description in flowery common language of the various chemical reactions which occur in a fire. The report was written for the benefit of the Michigan Department of Natural Resources and the news media. Therefore, it needed to be as descriptive as possible.

Having finally broken across the mental barriers which prevented me from thinking in terms of reactions occurring at such very low concentrations, I submitted this hypothesis to the Dioxin Task Force. Although it was understood and well received by most of the chemists in the group, some others, especially those responsible for waste combustion and coal-fired power generation, found it difficult to believe.

Determined to get unanimous agreement before the report was

released, we revised the document eleven times. Each revision was retyped and all copies of the previous version destroyed. Each adjective and adverb especially had to be approved by all 16 members of the group. All sentences and paragraphs were also approved in a similar manner.

Vigorous debates occurred throughout these revision sessions. After eight such revisions, I was getting rather tired of being involved in this discussion every day, seven days a week. I obviously needed a break – an opportunity to consider what we were saying from a perspective remote from the task force. I had thus arranged to visit friends and enjoy the Michigan State / Ohio State University football game in Columbus, Ohio, on October 23, 1978. At the Task Force meeting on October 20, Dr. Bumb announced that the Task Force would

Dr. Robert Bumb explains his feelings on learning the "Wizards" would analyze the samples and "Buck" Rogers would validate the data.

be meeting that Saturday and Sunday for four hours each day. I informed the group that I would be gone for the weekend.

"No, you are not!" I was told.

"Yes, I am!"

"Where do you propose to go?"

"To Columbus, Ohio, to the Ohio State / Michigan State football game."

"No, you're not! You'll surely tell someone about our discovery."

"No, I won't and yes I am going!" Frustrated, I picked up my chair and threw it across the room. Dr. Bumb needed to understand that science, not politics, was in charge.

"Yes, you are!" Bob Bumb agreed. "Have a good time. The Task Force will not meet this weekend."

With my good friend, Professor Thomas R. Sweet, of the Ohio State Chemistry Department, I discussed theories and hypotheses – especially the theory of evolution and possible proofs thereof. I came out of that discussion believing firmly in the "trace chemistry of fire" as a useful hypothesis.

Back in Midland we finished rewriting the report. Now we had to decide with whom and when we should share it.

The report had been written for the benefit of the Michigan Department of Natural Resources. It would have appeared reasonable to present them with a copy, a second copy going to the U. S. Environmental Protection Agency. However, that approach had been shown ineffective earlier when polybrominated biphenyls (PBBs), polychlorinated biphenyls (PCBs), and polychlorinated dioxins had been discovered in fish from the Tittabawassee River. The result was a premature selected leak to the press which seriously altered the meaning of the information in the report.

It, therefore, became obvious that the "trace chemistries of fire" report should be presented to the press directly. The Task Force, by a vote of 14 to 2 elected to hold a press conference simultaneously with presenting the report to the various concerned government agencies.

The press conference was held November 15, 1978. Professor Dr. Otto Hutzinger of the University of Amsterdam, the Netherlands, joined us. Dr. Bumb was the star of the show, as the reporters quickly perceived. Although several members of the Task Force were available, we were window dressing only.

The hypothesis was now public property!

References

1.C. Rappe, S. Markland, H. R. Buser, and H. P. Busshardt, "Formation of Polychlorinated Dibenzo-p-dioxin (PCDDs) and Dibenzofurans (PCDFs) by Burning or Heating Chlorophenates", Chemosphere, 7, 269 (1978).

2.H. G. Langer, T. P. Brady, and P. R. Briggs, "Formation of Dibenzo-p-dioxins and Other Condensation Products from Chlorinated Phenols and Derivatives", Environ. Health Persp., 5, 3 (1973).

3.J. W. Mellor, "A Comprehensive Treatise on Inorganic and Theoretical Chemistry", Volume II, "Chapter XVII The Halogens", p. 20 and 33-34, Longmans, Green and Co., London (1922).

4.G. Egloff, "Reactions of Paraffin Hydrocarbons", "Chapter II – The Reactions of Pure Hydrocarbons", p. 59-84 and 102-113, ACS Monograph Series, Reinhold Pub., Co., N.Y. (1937).

6.J. B. Edwards, "Combustion: The Formation and Emission of Trace Species", Ann Arbor Science, Ann Arbor, MI, 1977.

7.J. W. Larson, Ed., "Organic Chemistry of Coal", American Chemical Society, Washington, D.C., 1978.

8.H. B. Palmer and P. J. Seery, Amer. Rev. Phys. Chem., 24, 235, 1973.

9.National Academy of Sciences, "Chlorine and Hydrogen Chloride", Washington, D.C., 1976.

10.T. Kobayashi and T. Ikezawa, *Hyogo-Ken Kogai Kenkyusho Kenyu Hokoku*, 7, 1 (1975).

THE HYPOTHESIS

"... the streaming atoms
Fly on to clash together again, and make
Another and another state of things
Forever . . .", J. W. Miller, (about 1910).

1978

Simply stated, the trace chemistry of fire hypothesis recognizes that very many different chemical events take place in a fire of any kind. These numerous chemical events create numerous products. Most of these products are at a concentration level of a part per million or lower. Some of these products, such as the polynuclear aromatic hydrocarbons, have been known for decades. Others, such as the polychlorinated dibenzodioxins and dibenzodioxins at much lower concentrations were only discovered in combustion products after the analytical methodology to determine part per trillion levels was developed just before the hypothesis was announced. Many more have not yet been identified and reported. The number of different types of molecules which may possibly form on particulate matter approaches or exceeds that presently known to man.

All fuels are extremely complex mixtures of many chemical elements and compounds. When fuel is added to a hot chamber, a phenomena called "pyrolysis" immediately occurs. In this process, each component of the mixture is either degraded or condensed to form many other atoms and molecules. As these flow through the flame they are subjected to other conditions and collide with other particles resulting in yet other molecules, ions and free radicals being formed. Similar processes are repeated continually until the products are removed from the fire and returned to normal temperatures. In addition to the pyrolysis reactions; oxidation, reduction, and acidolysis occur.

The pyrolysis processes have been well known for many years[1] and have been reasonably well understood when working with sizeable amounts of pure materials under controlled conditions. However, reactions resulting in trace products at the part per million or lower level were not known. Modern analytical techniques are so incredibly sensitive and specific that results of chemical reactions occurring with very low yields, 0.0000000001 percent, can be identified and measured. From such analyses proof comes that the myriad of initial pyrolysis products formed in a fire are jumbled at low concentration in a sea of chemical reactivity. Within this sea; ions, electrons, free radicals, free atoms, and molecules form, collide, combine, and decompose. The process is repeated time and time again.

References

1. C. D. Hurd, "The Pyrolysis of Carbon Compounds", AACS Monograph Series, The Chemical Catalogue Company, N. Y. (1929).
2. R. R. Bumb, W. B. Crummett, S. S. Cutié, J. R. Gledhill, R. H. Hummel, R. O. Kagel, L. L. Lamparski, E. V. Luoma, D. L. Miller, T. J. Nestrick, L. A. Shadoff, R. H. Stehl, and J. S. Woods, "Trace Chemistries of Fire: A Source of Chlorinated Dioxins", Science, *210*, 385, (1980).

REACTION TO THE
HYPOTHESIS

"One observer will relate an event with the most extravagant
encomiums; another will detract from its real merit; a third
by some oblique insinuation will cast suspicion on the
motive; and a fourth will represent it as a crime of the
blackest dye.
These different descriptions represent the character of the
respective observers",
T. O Bergmann (1779).

1978

Many newspaper articles appeared as a result of the press conference. The headlines varied considerably in their meaning. Some
of those, which captured part of the meaning of the report, are:

11/16/78   "Dow Says Dioxins are Nothing New", Greenville, MI, Daily
            News

11/16/78   "Dow Says Dioxin Product of Fire", Ludington, MI, Daily News

11/16/78   "Dow Says Dioxin Chemical is Produced by Fire", Cheboygan,
            MI,   Daily Tribune

11/16/78   "Dow Says Dioxin Natural Fire Byproduct", Cadillac, MI,
            Evening   News

11/16/78   "Dow Says Chemical Present for Eons", Tri-Valley Herald

11/16/78   "Dow Says Fires Produce Dioxins", Monroe, MI, Evening News

11/16/78   "Dow Chemical Claims Dioxins Not New Thing", Hillsdale, MI,
            Daily News

11/16/78   "Fire Causes Dioxins", Port Huron, MI, Times Herald

11/16/78   "Dioxins Found in Smokestack", Oxnard, CA, Press Courier
11/16/78   "Dioxins as Old as Fire, Dow Announces", Iron Mountain, MI,
            News
12/15/87   "Dioxins Have Been Present Since the Advent of Fire, Dow
Says"   Science

Some headlines almost capture the meaning of the report, but extended the data to another issue not discussed in the report.

11/15/78   "Dioxin Found in Fish Everywhere, Dow Contends", Lansing,
MI,   State Journal
11/15/78   "Highly Toxic Dioxin Caused by Combustion", Petoskey, MI, News-
            Review
11/16/78   "Dow Chemical Co.: Combustion Produces Dioxin", Ironton, MI,
            Daily Globe
11/16/78   "Traces of Dioxin in Environment Said a Byproduct of Fire", Wash-
            ington Post
11/16/78   "Dow Claims Dioxin Test Shows Poison is as Old as Fire",
Muskegon,   MI, Chronicle
11/16/78   "Toxic Chemical Not Always Man-Made, Dow Study Finds",
Windsor,   Ont., Windsor Star
11/16/78   "Dioxin in Michigan Rivers May be From Dow Smokestack",
Corwallis,   OR, Gazette-Times
11/16/78   "Toxic Chemical Traces May be Result of Nature", Coldwater, MI,
            Daily Reporter

Other headlines were highly sensationalized by the use of scary adjectives.
11/16/78   "Fire Called Source of Deadly Dioxins", The Seattle Times

11/16/78   "Dow Sees Self as Source of Deadly, Common Dioxin" Lansing,
MI,   State Journal

11/16/78   "Dow Says Burning Produces Deadly Dioxins", Flint, MI, Journal

11/16/78   "Deadly Dioxins Wherever Fire", Houghton, MI, Daily Mining

Ga-   zette
11/16/78   "Deadly Dioxins are Nothing New: Dow", Sault Ste. Marie, MI,
            Evening News
11/16/78   "Dow: Deadly Dioxin is Common", Sault Ste. Marie, MI, Evening
            News
11/16/78   "Deadly Substance Traced to Fires", Oakland, CA, Tribune
11/16/78   "Deadly Chemical 'Natural, Common'", Santa Ana, CA, Register

A few headlines were very misleading and even false.

11/16/78   "Nature's Deadly Chemical. Firm Finds Viet Defoliant is Fire
            Byproduct", Newark, NJ, The Star Ledger
11/16/78   "Poison Blamed on Natural Causes"   San Diego, CA, Union
11/16/78   "Dow Denies a Role in Fish Poisonings", New York Times
11/16/78   "Deadly Dioxins Pose No Threat to Humans", Petoskey, MI, News-
            Review
11/16/78   "Firm Exudes Deadly Chemical", Pasadena, CA, Star News
11/16/78   "Deadly Toxins Aren't Hazardous", Oceanside, CA, Blade Tribune

Some headlines were looking to fuel the issue.

11/17/78   "Scientists Skeptical About Dow Claim", Iron Mountain, MI, News
11/17/78   "Skeptical of Dow", Ludington, MI, Daily News
11/17/78   "Magazine Takes Issue with Dow", Midland, MI, Midland Daily
            News

Only one headline suggested that the work may be of scientific
merit.

11/22/78   "Dow Study Could be Important Breakthrough", Midland, MI, Mid-
            land Daily News

Over the next several years, the story of the "trace chemistries of
fire" can be appreciated by reading selected headlines and titles of
articles that pertain to the subject.

01/1979 "Ubiquitous Dioxins Claimed to be Normal, Combustion Products",
   Pollution Engineering
01/1979 "Dioxins, Dioxins Everywhere", Chemistry
01/12/79, "Dow Finds Support, Doubt for Dioxin Ideas", C&E News
02/1979 "Dioxins Occur Everywhere", Water Well Journal
02/16/79, "EPA Pollution Report Contradicts Dow Study", Saginaw, MI, Saginaw
   News
02/16//9, "Taint Blamed on Dow", Jackson, MI, Citizen Patriot
02/17/79, "Dow Chemical Co. Responds to EPA Report on Poison", Monroe, MI,
   Evening News
02/19/79, "EPA Fingers Dow Chemical Plant as Source of Michigan Pollution",
   Air/Water Pollution Report
02/17/79, "Dow Debates U. S. on Poison", The Detroit News
02/17/79, "Dow Admits Pollution, Contests Source", Saginaw, MI, Saginaw News
02/16/79, "Dow, EPA to Discuss Dioxins Found in River", Bay City, MI, The Bay
   City Times
03/14/79, "DNR Disputes Dow Claim", Bay City, MI, The Bay City Times
08/17/79, "Contaminant Dioxin is Found in Coal Plant Emissions", Washington
   Post
08/16/79, "Dioxin, The Poison in Herbicide, Found in Coal Emissions", Day
09/1980 "Dioxins from Garbage", F. W. Karasek, Canadian Research
02/19/81, "Dioxin Haunts EPA Refuse Plants", ENR
09/02/81, "Dow Tail Wagged Watchdog", Saginaw, MI, Saginaw News
05/1986 "Dioxin Danger from Garbage Incineration", F. W. Karasek and Otto
   Hutzinger, Anal. Chem., 58, 633A
04/24/-86, "Swedish Program Alters Opinion on Dioxins", H. Almstroem Warner,
   Swedish Correspondent.

We sent copies of the report to a few of the leading scientists in
the United States. Comments were generally favorable, as can be seen
from the quotations listed below. Generally, physical and analytical
chemists supported the hypothesis; biologists gave no opinion, but
found it motivating. On the other hand, organic synthesis chemists
found the concepts difficult and believed the data to be inadequate.

Comments from Some Leading Scientists

"I have read your report, and am most favorably impressed with the plan of the work, with the resources that have been put to the investigation, with the very good English and careful choice of words used in the writing, with the conservative attitude displayed in the report, and with the confidence displayed in the results so far presented. In short, it is a very good work that should appeal to any level-headed, unbiased reader." Frederick D. Rossini, Chairman Emeritus, Department of Chemistry, Rice University, December 2, 1978.

"I concur with your conclusions concerning the ubiquity of these materials." Joseph F. Barzelleca, Professor of Pharmacology, Virginia Commonwealth University, December 14 , 1978.

"It seems to me that you people have opened a whole new field of great interest. You use the word "ubiquitous" to describe these dioxins, and that seems highly appropriate at this point." William C. Ackerman, Chief, Illinois State Water Survey, December 8, 1978.

"This is one of the most exciting development, informative, and I believe, useful studies that I have seen in a long time." Emil M. Mrak, Chancellor Emeritus, University of California, December 6, 1978.

"I should think that this paper will stand as a revolutionary milestone introducing a new era in considering environmental problems. My congratulations to Dow for a superb job." H. E. Carter, Arizona Medical Center, University of Arizona, December 5, 1978.

"The report concludes that chlorinated dioxins are formed in many combustion processes and implies that they are the inevitable products of the trace chemistry of fire. The data bearing on this point are grossly inadequate . . ." Andres S. Kender, Organic Chemistry Professor, University of Rochester, November 27, 1978.

"In general the data presented in this report do not support conclusions expressed therein that chlorinated dioxins are the ubiquitous trace products of combustion." Howard A. Tanner, Director, Department of Natural Resources, State of Michigan, November 5, 1979.

"Since all combustions produce small quantities of dioxin, Dow wants to put the responsibility for such pollution on everyone, rather than assume it themselves." Ray Meluch, Great Lakes Greenpeace, 1979.

"Polychlorinated dibenzo-p-dioxins (PCDD) and polychlorinated dibenzofurans are ubiquitous environmental contaminants." Richard E. Tucker, Alvin L. Young, and Allen P. Gray, Editors, "Human and Environmental Risks of Chlorinated Dioxins and Related Compounds", Plenum Press, 1981.

Statements About the Hypothesis by Dioxin Scientists

"In view of the recent knowledge on the *de novo* formation of aromatic compounds from organic material such as polyethylene and inorganic chloride under pyrolytic conditions, however, a thermal synthesis of chlorodibenzo-p-dioxins, chlorodibenzofurans or, indeed, chlorophenols is entirely possible."
—O. Hutzinger, et al., *1977*

"We now think dioxins have been with us since the advent of fire."
—Robert Bumb, *1978*

"As, for example, forest fires have occurred throughout history, the pollution of chlorinated hydrocarbons is also a pre-industrial phenomena."
—B. Ahling, *1981*

"Dioxin—and furan-formation as part of the general *de novo* for-

mation of aromatic hydrocarbons seems to be a well established fact in flame chemistry."
                                        —K. Ballschmiter, et al., *1986*

"We have now proven *de novo* synthesis of dioxins."
                                        —H. Hagenmeier, *1986*

"PCDD's and PCDF's must have been around since the first pre-historic forest fire that ever occurred."
                                        —Sidney Rankin, *1986*

"As polychlorinated dibenzodioxins and dibenzofurans are to be considered ubiquitous in technologically advanced countries, the 'Trace Chemistries of Fire' hypothesis has been established in the late seventies."
                                        —O. Hutzinger, *1988*

Although the "trace chemistries of fire" hypothesis has not been proven (nor can it be), it is now a part of the consciousness of the scientific community and thus is a powerful force for further research and will be taking a continual role in the consideration of the quality of the environment.

DEFENSE OF THE WORK

*"One cannot judge where information is appropriate, or
necessary, or sufficient, or excessive except in terms of the USES to
which it will be put",
George Zissis, Environ. Res. Inst. of Michigan.*

1979-1983

By the middle of 1979, the Dow Dioxin Task Force had been
totally disbanded – most members having been promoted into other
departments, some remote from the Midland situation. Only Les
Lamparski, Terry Nestrick, Don Townsend and myself remained in
positions to defend, explain, and further develop our discoveries of
the "trace chemistries of fire" hypothesis. We did this even as the
manuscript was being considered for publication in Science. In addition to telling our story we needed to learn how other scientists re-
acted to the work, how they might use the data, and what, if any,
more data was needed to prove or disprove the notion that dioxins
were ubiquitous.

The hypothesis was presented and discussed at many scientific
conferences. The American Oil Chemists Society was the first to invite us. Don Townsend and I were asked to speak at the symposium
with the theme "Potential Toxicants Related to Fats and Oils and
Their Derivatives in Commercial Processing" on Tuesday, March 20,
1979, in Elizabeth, New Jersey. The very first speaker, the chief chemist
from one of the largest bakeries in the United States, surprised us by
referring to newspaper articles from the New York Times and Wall
Street Journal as the best sources of scientific information. He told us
that he believed his mission of the day to be to warn oil chemists
about toxic chemicals, which may be found in various oils used in

food. "One of these of greatest concern," he said, "is dioxin, which has now been found in corn oil." We were shocked by this statement. All the scientific data we had seen showed that 2,3,7,8-TCDD was not translocated in plants and thus did not get into the seed or grains of plants.

The audience gasped. The speaker appeared encouraged by this response and he amplified his statement. He declared that dioxin migrated through roots of the corn plant, up the stalk, and into the ears of corn where it was concentrated in the oil of the kernels.

We were aghast! Such nonsense! But, how could we challenge him without risking our own credibility? We decided to wait until after we had given our talk. However, we need not have been concerned. A hand was raised in the back of the room and an unfamiliar, lined, weathered, leathery face framed in gray said, "I don't believe dioxin would behave that way!"

"It most certainly does", the speaker replied.

"What evidence do you have?"

"It's spelled out in this newspaper article! I have a copy right here."

"Would you read it, please?"

"Certainly!" He then read an item to the effect that an increase in tumors was observed in animals fed corn oil spiked with *dioxane*. The audience laughed. *Dioxane* is a compound which is completely different from *dioxin*. Our speaker had confused the two.

Other societies that invited us to speak on the subject and the dates at which one or more of us appeared, included the following:

Scientific Societies
- American Chemical Society National Meeting, Washington, D.C., September 13, 1979.
- American Chemical Society, Indianapolis, Indiana Section, Butler University, October 25, 1979.
- American Chemical Society, Cincinnati, Ohio Section, January 9, 1980.
- American Chemical Society, Buffalo, New York Section, March 16, 1983.

- American Chemical Society, Virginia Section, October 18, 1985.
- Collaborative International Pesticides Advisory Council, Baltimore, Maryland, June 7, 1979.
- Royal Chemical Society, Sussex, England, April 1980.
- International Association of Environmental Analytical Chemists.
  International Society of Toxicology & Environmental Chemists, Instituto Superiore di Sanita, Rome, Italy, October 22, 1980.
- International Union of Pure and Applied Chemistry, Kyoto, Japan, August 27, 1982.
- International Union of Pure and Applied Chemistry, Jekyll Island, Georgia., May 7, 1979.
- American Society of Mechanical Engineers Research Committee, McLean, Virginia, January 24, 1984.
- Association of Official Analytical Chemists, Washington, D.C., October 29, 1984.

Academic Institutions
- University of Arizona, Tucson, Arizona, September 1979.
- Arizona State University, Tempe, Arizona, September 1979.
- University of New Mexico, Albuquerque, New Mexico, September 1979.
- University of Georgia, Atlanta, Georgia, May 9, 1979.
- Michigan Technological University, Houghton, Michigan, April 1, 1981.
- Ohio State University, January 27, 1981.
- University of Wisconsin, April 23, 1981.
- Lawrence Institute of Technology, Detroit, Michigan, March 24, 1980.
- University of Amsterdam, October 13, 1980.
- University of Massachusetts, Amherst, Maine, May 5, 1981.
- University of Manitoba, Winnipeg, Canada, April 1980.
- Virginia Polytechnic Institute, October 10, 1983.
- State University of New York, Buffalo, New York, March 17,

1983.

- Michigan State University, Lansing, Michigan, December 6, 1983.

Government Agencies

- Government of Canada Contaminants Control Branch, Toronto, Canada, February 13, 1980.
- State of New York Department of Health, Albany, New York, April 21, 1980.
- State of Michigan Department of Natural Resources, Lansing, Michigan, September 3, 1980.
- State of Michigan Department of Natural Resources, Lansing, Michigan, February 2, 1981.
- National Research Council of Canada, Toronto, February 1980.
- National Research Council of Canada, Ottawa, July 1-3, 1980.
- U. S. Department of Agriculture, Washington, D.C., November 30, 1978.
- National Bureau of Standards, Bethesda, Maryland, October 25, 1982.
- Health and Welfare Canada / Environment Canada Expert Advisory Committee on Dioxins, Ottawa, Canada, 1982.
- Machida Cultural Centre, Machida, Japan, September 12, 1986.

The U. S. Environmental Protection Agency is notably missing from this list. Although I had served on the EPA Science Advisory Board, I was now ostracized by them.

TESTING THE HYPOTHESIS

"The scientific value of thoroughly sound hypotheses is enhanced
daily both by known facts that they are constantly assimilating
and new facts that they are continually revealing",
J. Ward (1899).

1978-1989

Immediately after "the trace chemistries of fire" report was released we received suggestions from various eminent scientists in the
United States for experiments designed to test the hypothesis. These
were often accompanied by samples which, if found to contain the
chlorinated dioxins or other chlorinated organic matter at levels of a
part per quadrillion, would prove that chlorinated dioxins are ubiquitous. These included:

1. Drill cores high in carbonaceous matter taken at depths
    corresponding to 5000, 12000, and 35000 years from
    under ancient lake beds in Israel.
2. Condensates from sea breezes taken on remote islands in the
    Pacific where fire is not present.
3. Ice taken from the center of an ancient glacier.
4. Ash taken from the mouth of a volcano.
5. A fossilized fish dated as 12,000 years old.

We did not analyze any of these samples. They were probably
contaminated while they were collected or transported. They had
been collected appropriately for other analyses, but not for the determination of dioxins. Furthermore, we were also concerned about how

we could open the sample containers without contaminating the samples at these very low levels.

Two other suggestions from the scientists were easier to carry out experimentally. These were:

1. Determine if chlorinated dioxins can be found by burning fossil fuel in the presence of chlorine or inorganic chloride.
2. Determine if chlorinated dioxins are present in fish taken from rivers remote from pesticide manufacturing facilities, but close to incinerators and fossil-fuel powerhouses.

We did some exploratory work in both of these determinations and found positive results in both cases. These experiments gave us further confidence in the hypothesis. Since those early experiments many investigations which test the hypothesis have been made. These may be listed as a number of different types, as follows:

1. Studies on the combustion products of incinerators. Emissions, usually fly ash, have been examined from municipal incinerators in the Netherlands, Sweden, Switzerland, France, Canada, Italy, Japan, West German, Belgium and the United States. All municipal incinerators thus far examined produce emissions which contain part per trillion levels of the chlorinated dibenzodioxins. We, therefore, conclude that this is a general phenomenon.
2. Studies on the combustion products of wood. Soot from the burning of wood has been studied in Sweden, the Netherlands, the United States, and Canada. All studies show the presence of part per trillion levels of the chlorinated dioxins. We now conclude that this is a general phenomenon.
3. Studies on the combustion products of light hydrocarbons. Recent finding of the PCDDs in soot from the tailpipes of automobiles in Germany and Sweden together with earlier qualitative data taken in the United States suggests that this too may be a general phenomenon.

4. Studies on the combustion products of coal. The presence or absence of PCDDs and their precursors in emissions from coal combustion has been differently reported and debated by many investigators. About half of the studies report their presence and about half do not find them. Unfortunately, many of the studies include only one or two samples. This is an insufficient number to characterize the coal combustion process. Until coal-fired powerhouses are studied in detail using the most specific analytical methodology, conclusion on chlorinated dioxin formation in coal combustion can only be made by extrapolation. Therefore, definitive statements cannot be made.

5. Studies to identify precursors and elucidate reaction mechanisms. Precursors for the formation of the PCDDs are continually being identified. The list is no longer limited to the chlorinated phenols and their derivatives. However, for purposes of testing the "trace chemistries of fire" hypothesis, the list of identified organic emissions from selected combustion processes developed by Junk and Ford is most significant. They found 331 different components in total: 109 from the combustion of coal, 235 from the incineration of wastes, and 69 from coal/refuse combustion processes. With the more sensitive analytical methods in use today, the number that could be identified in the same samples would be expected to have increased by at least ten times.

6. Theoretical studies related to the "trace chemistries of fire" hypothesis. Research on the thermodynamics, Kinetics, and mechanisms of formation of PCDDs and PCDFs has continued. Results of these studies tend to support the hypothesis.

Selected References

Municipal Incinerators

1.J. W. A. Lusenhouwer, K. Olie, and O. Hutzinger, "Chlorinated Dibenzo-p-dioxins and Related Compounds in Incinerator Effluents", Chemosphere, *9*, 501 (1980).

2. Cavallaro, G. Bandi, G. Invernizzi, L. Luciani, E. Mongini, and A. Garni, "Sampling, Occurrence and Evaluation of PCDDs and PCDFs from Incinerator Solid Urban Waste", Chemosphere, 9, 611 (1980).

3. G. A. Eiceman, R. E. Clement, and F. W. Karasek, "Analysis of Fly Ash from Municipal Incinerators for Trace Organic Compounds", Anal. Chem., 51, 2243 (1979).

4. T. O. Tiernan, M. L. Taylor, J. H. Garrett, G. f. Van Ness, J. G. Solch, D. A. Deis, and D. J. Wagel, "Chlorodibenzodioxins, Chlorodibenzofurans and Related Compounds in the Effluents from Combustion Processes", Chemosphere, 12, 595, (1983).

5. K. Ballschmiter, W. Zaller, C. Scholz, and A. Nollrodt, "Occurrence and Absence of Polychloro-dibenzofurans and Polychloro-dibenzodioxins in Fly Ash from Municipal Incinerators", Chemosphere, 12, 585, (1983).

6. De Fre, "Dioxin Levels in the Emissions of Belgian Municipal Incinerators", Chemosphere, 15, 1255 (1986).

7. E. Besofenati, R. Pastorelli, H. G. Castelli, R. Fanelli, A. Carminati, A. Farneti, and M. Lodi, "Studies on the tetrachloro-dibenzo-p-dioxins (TCDD) and tetrachlorodibenzo-furans (TCDF) emitted from an urban incinerator", Chemosphere, 15, 557, 1986.

8. H. Y. Tong and F. W. Karasek, "Comparison of PCDD and PCDF in Fly Ash Collected from Municipal Incinerations of Different Countries", Chemosphere, 15, 1219, 1986.

9. K. Olie, J. W. A. Lustenhouwer and O. Hutzinger, "Polychlorinated Dibenzo-p-dioxins and Related Compounds in Incinerator Effluents", "Chlorinated Dioxins and Related Compounds", O. Hutzinger, R. W. Frei, E. Merian, and F. Pocchiari, Editors. Pergamon Press, Oxford, 1982, p 227.

10. Carvallaro, L. Luciana, G. Ceroni, I. Rocchi, G. Invernizzi, and A. Garin, "Summary of Results of PCDDs Analyses from Incinerator Effluents", Chemosphere, 11, 859 (1982).

11. K. W. Wang, D. H. Chiu, P, K. Leung, R. S. Thomas and R. C. Lao, "Sampling and Analytical Methodologies for PCDDs and PCDFs in Incinerators and Wood Burning Facilities", "Human and Environmental Risks of Chlorinated Dioxins and Related Compounds", R. E. Tucker, A. L. Young, and A. P. Gray, Editors. Plenum Press, 1982, p. 113.

12. C. Chiu, R. S. Thomas, J. Lockwood, K. Li, R. Halenan, and R. C. Lao, "Poly-

chlorinated Hydrocarbons from Power Plants and Municipal Incinerators", Chemosphere, *12*, 607 (1983).

13. U. Samuelson and A. Lindskog, "Chlorinated Compounds in Emissions From Municipal Incineration", Chemosphere, *12*, 665, (1983).

14. G. Eklund and B. Stromberg, "Detection of Polychlorinated Polynuclear Aromatics in Flue Gases from Coal Combustion and Refuse Incinerators", Chemosphere, *12*, 657, (1983).

15. C. Rappe, H. R. Buser, and H. P. Bosshardt, "Polychlorinated Dibenzo-p-dioxins (PCDDs) and Dibenzofurans (PCDFs): Occurrence, Formation, and Analysis of Environmentally Hazardous Compounds", CIPAC Proceedings Symposium Papers, W. R. Bontoyan, editor, CIPAC Publications (1979), p 74.

16. H. R. Buser and C. Rappe, "High Resolution Gas Chromatography of the 22 Tetrachlorodibenzo-p-dioxin Isomers", Anal. Chem., *52*, 2262 (1980).

17. H. R. Buser and C. Rappe, "Identification of Substitution in Polychlorinated Dibenzo-p-dioxins (PCDDs) by Mass Spectrometry", Chemosphere, *7*(1978).

18. J. Janssens, L. Van Vaeok, P. Schepens, and F. Adams, "Qualitative and Quantitative Analysis of Emissions of a Municipal Incineration Installation", Comm. Env. Communities, EUR7624, 28 (1982).

19. A. Liberti and D. Brocco, "Formulation of Polychlorodibenzodioxins and Polychlorodibenzofurans in Urban Incinerators Emission", "Chlorinated Dioxins and Related Compounds", O. Hutzinger, R. W. Frei, E. Merian, and F. Pocchiari, editors. Pergamon Press, Oxford, 1982, p 245.

20. A. Liberti, P. Ciccioli, E. Brancaleoni, and A. Cecinato, "Determination of Polychlorodibenzo-p-dioxins and Polychloinate-p-dibenzofurns in Environmental Samples by Gas Chromatography/Mass Spectrometry", J. Chromat., *242*, 111 (1982).

21. L. Masselli, F. Rifiro, and B. Villori, "Analysis of the Effluents of an Urban-Solid Refuse Incinerator: Study of Methods of Extraction and Analysis for the Quantitative Determination of Polychlorodibenzo-p-dioxins", Anal. di Chem., p 557 (1981).

22. F. Gizzi, R. Reginato, E. Benfenati, and R. Fanelli, "Polychlorinated Dibenzo-p-dioxins (PCDD) and Polychlorinated Dibenzofurans (PCDF) in Emissions from an Urban Incinerator 1. Average and Peak Values", Chemosphere, *11*, 577 (1982).

23. G. A. Eiceman, R. E. Clement, and F. W. Karasek, "Variations in Concentration

of Organic Compounds Including Polychlorinated Dibenzo-p-dioxins and Polynuclear Aromatic Hydrocarbons in Fly Ash from a Municipal Incinerator", Anal. Chem., 53, 955 (1981).

24. F. W. Karasek, R. E. Clement and A. C. Viau, "Distribution of PCDDs and Other Toxic Compounds Generated on Fly Ash Particulates in Municipal Incinerators", J. of Chromatog., 239, 173 (1982).

25. D. P. Redford, C. L. Haile, and R. M. Lucas, "Emissions of PCDDs and PDCFs from Combustion Sources", Human and Environmental Risks of Chlorinated Dioxins and Related Compounds", R. E. Tucker, A. L. Young, and A. P. Gray, editors. Plenum Press, 1982, p 143.

26. H. R. Buser and H. P. Bosshardt, "Polychlorierte Dibenzo-p-dixoine, Debenzofurance, and Benzole in der Asche Kummunaler and indusrieller Verbrennungsanlagen", Mitt. Gebiete Lebensm, Hyg., 69, 191, (1978).

27. D. W. Hryharczuk, W. A. Withrow, C. S. Hesse, and V. R. Beasley, "A Wire Reclamation Incinerator as a Source of Environmental Contamination with Tetrachlorodibenzo-p-dioxins and Tetrachlorodibenzofurans", Arch. Environ. Health, 36, 228 (1981).

28. C. Rappe, S. Marklund, P. A. Bergguist, M. Hannson, "Polychlorinated Dibenzo-p-dioxins, Dibenzofurans and Other Polynuclear Aromatics Formed During Incineration and Polychlorinated Biphenyl Fires", in Chlorinated Dioxins and Dibeazofurans in the Total Environment, G. Choudhry, L. H. Keith, and C. Rappe (eds.), Ann Arbor Science, Butterworth Publ., Woburn, MA, 1983, p 99-124.

29. H. Vogg and L. Stieglitz, "Thermal Behavior of PCDD in Fly Ash from Municipal Incinerators", Chemosphere, 15, 1373 (1986).

30. W. M. Shaub, "Technical Issues Concerned with PCDD and PCDF Formation and Destruction in MSW Fired Incinerators", U. S. Department of Commerce N.B.S. Reports, NBSIR 84-2975 (1984).

31. W. M. Shaub, W. Tsang, "Overview of Dioxin Formation in Gas and Solid Phases Under Municipal Incineration Conditions" in: Chlorinated Dioxins and Dibenzofurans in the Total Environment, II. Ed. L. H. Keith, C. Rappe, G. Choudhry, Butterworth Publishers, 1985, Stoneham, MA).

32. M. F. Gonnord, F. W. Karasek, and C. Finet, "Formation of Dioxin and Chlorobenzenes from PVC in a Municipal Incinerator", "(Formation de dioxines et de chlorobenzenes a partir de PVC dans un incinerateur de dechets urgains)", T.S.M.-1, 'Eau., May 1985, Vol. 80, No. 5, p 211-216.

33.E. Benfenati, F. Gizzi, R. Reginato, R. Fanelli, M. Lodi, and R. Tagliaferri, "PCDDs and PCDFs in Emissions from an Urban Incinerator. 2. Correlations Between Concentration of Micropollutats and Combustion Conditions", Chemosphere, *12*, 1151-1157 (1983).

34.S. J. Thorndyke, "Determination of Chlorinated Dibenzo-p-dioxins, Chlorinated Dibenzofurans, Chlorinated Biphenyls, Chlorobenzenes and Chlorophenols in Air Emissions and Other Process Streams at SWARU in Hamilton", Report No. ARB-02-84-ETRD. Part 1, Sampling –4318/G, Revised, Ministry of the Environment, Toronto, 1983).

35.B. Commoner, M. McNamara, K. Shapiro, T. Webster, *Environmental and Economic Analysis of Alternative Municipal Solid Waste Disposal Technologies. II. The Origins of Chlorinated Dioxins and Dibenzofurans Emitted by Incinerators that Burn Unseparted Municipal Solid Waste, and An Assessment of Methods of Controlling Them.* Center for the Biology of Natural Systems Publication, Queens College, Flushing, N.Y., December 1, 1984.

36.Envirocon Ltd., *Report on Combustion Testing Program at the SWARU Plant, Hamilton-Wentworth*, prepared for Ontario Ministry of the Environment Air Resources Branch, Report # ARB-43-84-ETRD, January 1984).

37.M. F. Gonnord, G. Gueochon, A. C. Viau, F. W. Karasek, and C. Finet, "Survey and Control of Organic Halide (PCDD) Emissions in Incinerator Fly Ash", Doc. Eir. Abwasser Abfall Symposium 6th, p 669-686 (1984).

38.F. W. Karasek, H. Y. Tong, D. L Shore, P. Helland and E. Jellum, "Identification of Organic from Incineration of Municipal Waste by High-Performance Liquid Chromatographic Fractionation and Gas Chromatography – Mass Spectrometry", J. Chromato., *285*, 423-441, 1984.

39.W. Niessen, "Production of polychlorinated dibenzo-p-dioxins (PCDD) and Dibenzofurans (PCDF) from Resource Recovery Facilities. Part II", Proceedings of the 1984 National Waste Processing Conference, ASME, New York, p 358, 1984.

40.V. Ozvacic; G. Wong, H. Tosine, R. Clement, J. Osbrorne, and S. Thorndyke, "Determination of chlorinated dibenzo-p-dioxins, chlorinated dibenzofurans, chlorinated biphenyls, chlorobenzenes and chlorophenols in air emissions and other process steams at SWARU in Hamilton, Ontario Ministry of Environment Report ARB-02-84-ETRD, Toronto, Ontario, Canada, July 1984.

41.R. E. Clement, H. M. Tosine, J. Osborne, V. Ozvacic, and G. Wong, "Levels of Chlorinated Organics in a Municipal Incinerator", in: *Chlorinated Dioxins* and *Dibenzofurans in the Total Environment, II*, eds., L. H. Keith, C. Rappe, G. Choudhry, Butterworth Publishers 1985, Stoneham, Maine, pp 489-514.

42.M. Hovaguchi, H. Ogawa, K. Ose, S. Tomisawa, and S. Matsuura, "PCDDs and PCDFs from the MSW Incinerator", Chemosphere, 18, 1785 (1989).

43.W. Schafer and K. Ballschmiter, "Monobromo-polychloro-derivatives of Benzene, Biphenyl, Dibenzofurans and Dibenzodioxins formed in Chemical Waste Burning", Chemosphere, *15*, 755, 1986.

44.J. Johnson, "Incinerators Targeted by EPA", Environ. Sci. Technology, *29*, 33A, 1995).

Combustion of Wood

1.B. Ahling and A. Lindskog, "Emission of Organic Substances from Combustion", "Chlorinated Dioxins and Related Compounds", O. Hutzinger, R. W. Frei, E. merian, and F. Pocchiarai, editors. Pergamon Press, Oxford, 1982, p 215.

2.T. O. Tiernan, "Direct Testimony of Dr. Thomas O. Tiernan. Before the Environmental protection Agency of the United States of America", in Re: The Dow Chemical Company, et al. Docket Nos. 415, et al. Exhibit No. 222.

3.National Research Council, Canada, "Polychlorinated Dibenzo-p-dioxins: Criteria for The effects on Man and His Environment", NRCC No. 18574, p 24-36.

4.T. J. Nestrick and L. L. Lamparski, "Assessment of Chlorinated Dibenzo-p-dioxin Formation and Potential Emission to the Environment from Wood Combustion", Chemsophere, *12*, 617, 1983.

5.T. J. Nestrick and L. L. Lamparski, "Isomer-Specific Determination of Chlorinated Dioxins for Assessment of Formation and Potential Environmental Emission from Wood Combustion", Anal. Chem., *54*, 2292, 1982.

6.D. K. W. Wang, D. H. Chiu, P. K. Leung, R S. Tomas and R. C. Lao, "Sampling and Analytical Methodologies for PCDDs and PCDFs in Incinerators and Wood Burning Facilities", "Human and Environmental Risks of Chlorinated Dioxins and Related Compounds", R. E. Tucker, A. L. Young, and A. P. Gray, editors. Plenum Press, 1982, p 113.

7.R. E. Clement, H. M. Tosine, and B. Ali, "Levels of Polychlorinated Dibenzo-p-dioxins and Dibenzofurans in Wood Burning Stoves and Fireplaces", Chemosphere, *14*, p 815-819, 1985.

Combustion of Light Hydrocarbons

1.S. Kirshman and R. A. Hites, "Chlorinated Organic Compounds Formed in a Methane – Dichloromethane Flame", Chemosphere, *9*, 679, 1980.

2.H. Svec, private communication

3.T. O. Tiernan, "Direct Testimony of Dr. Thomas O. Tiermnan. Before the Environmental Protection Agency of the United States of America", in Re: The Dow Chemical Company, et al. Docket Nos. 415, et al. Exhibit No. 222.

4.R. R. Bumb, W. B. Crummett, S. S. Cutié, J. R. Gledhill, R. H. Hummel, R. O. Kagel, L. L. Lamparski, E. V. Luoma, D. L. Miller, T. J. Nestrick, L. A. Shadoff, R. H. Stehl, and J. S. Woods, "Trace Chemistries of Fire: A Source of Chlorinated Dioxins", Science, *210*, 385, 1980.

5.K. Ballschmiter, H. Buchert, R. Niemczyk, A. Munder, and M. Swerev, "Automobile Exhausts versus Municipal-waste Incineration as Sources of the Polychloro-debenzodioxin (PCDD) and Furans (PCDF) Found in the Environment", Chemosphere, *15*, 901, 1986.

6.S. Marklund, C. Rappe, M. Tysklind, and K. Egeback, "Identification of Polychlorinated Dibenzofurans and Dioxins in Exhausts from Cars run on Leaded Gasoline", Chemosphere, *16*, 29, 1987.

7.H. Thomas, "PCDD/F Concentrations in Chimney Soot from House Heating Systems," Chemosphere, *17*, 1369, 1988.

Coal-Fired Powerhouses

1.B. J. Kimble, and M. L. Gross, "Tetrachlorodibenzo-p-dioxin Quantitation in Stack Collected Fly Ash", Science, *207*, 59, 1980.

2.C. R. Chiu, R. S. Thomas, J. Lockwood, K. Li, R. Halenan, and R. C. C. Lao, "Polychlorinated Hydrocarbons from Power Plants, Wood Burning and Municipal Icinerators", Chemosphere, 12, *607*, 1983.

3.J. M. Czuczwa, "The Environmental Fate of Combustion Generated Polychlorinated Dibenzo-p-dioxins and Dibenzo-furans", Doctoral Thesis, Indiana University, 1984. University Microfilms International, 1984, Ann Arbor, Michigan.

4.F. L. DeRoos and a. Bjorseth, "TCDD Analysis of Fly Ash Sample". Report prepared by Battelle Columbus Laboratories for the U. S. Environmental Protection Agency, June 15, 1979.

5.G. Eklund and B. Stromberg, "Detection of Polychlorinated Polynuclear Aromatics in Flue Gases from Coal Combustion and Refuse Incinerators", Chemosphere, 12, 657, 1983.

R. L. Harless and R. G. Lewis, "Quantitative Determination of 2,3,7,8-Tetrachlorodibenzo-p-dioxin Residues by Gas Chromography / Mass Spectrometry", *Impact of Chlorinated Dioxins and Related Compounds in the Environment*. O. Hutzinger, R. W. Frei, E. Marian, and F. Pocchiari, editors. Pergamon Press, New York, p 25-35, 1982.

7.D. P. Redford, C. L. Haile, and R. M. Lucas, "Emissions of PCDDs and PCDFs from Combustion sources", in *Human and Environ. Risks of Chlorinated Dioxins and Related Compounds*, p 143-152, R. E. Tucker, A. L. Young, and A. P. Grey, editors. Plenum Press, New York, p 143-152, 1983.

Precursors

1.R. Lindahl, C. Rappe, and H. R. Buser, "Formation of Polychlorinated Dibenzofurans (PCDFs) and Polychlorinated Dibenzo-p-dioxins (PCDDs) from the Pyrolysis of Polychlorinated Diphenyl Ethers", Chemospere, 9, 351, 1980.

2.D. Brocco, a. Cecinato, and A. Liberti, "Polychlorodibenzodioxins in the Environment", Chemosphere, 7, 199, 1978.

3.A. Liberti and D. Brocco, "Formation of Polychlorodibenzodioxins and Polychlorodibenzofurans in Urban Incinerators Emissions", "Chlorinated Dioxins and Related Compounds", O. Hutzinger, R. W. Frei, E. Merian, and F. Pocchiari, editors. Pergamon Press, Oxford, p. 245, 1982.

4.K. M. Olie, V. D. Berg, and O. Hutzinger, "Formation and Fate of PCDD and PCDF from Combustion Processes", Chemosphere, 12, 627, 1983.

5.K. Ballschmiter, W. Zaller, C. Scholz, and A. Nollrodt, "Occurrence and Absence of Polychloro-dibenzofurans and Polychloro-dibenzodioxins in Fly Ash from Municipal Incinerators", Chemosphere, 12, 585, 1983.

6.G. A. Junk and C. S. Ford, "A Review of Organic Emissions from Selected Combustion Processes", Chemosphere, 9, 187, 1984.

7.K. Ballschmiter, L. Braunsmiller, K. Niemczyk, and M. Severev, "Chlorobenzenenes and Chlorophenols as Pecursors in the Formating of Polychloro-dibenzodioxins and dibenzofurans in Flame Chemistry", Chemosphere, 17, 995, 1988.

Theoretical Studies

1.G. Choudhary, K. Olie, and O. Hutzinger, "Mechanisms in the Thermal Formation of Polychlorinated Dibenzo-p-dioxins and Related Compounds", "Chlorinated Dioxins and Related Compounds", O, Hutzinger, R. W. Frei, E. Merian, and F. Pocchiari, editors. Pergamon Press, Oxford, p 275, 1982.

2.C. Shih, D. Ackerman L. Scinto, and B. Johnson, "Emissions of Polychlorinated Dibenzo-p-dioxins (PCDDs) and Dibenzofurans (PCDFs) from the Combustion of Fossil Fuels, Wood and Coal-Refuse". Draft, EPA Contract No. 68-01-3138, *TRW* Environmental Engineering Division, December 1980.

3.R. Lindahl, C. Rappe, and H. R. Buser, "Formation of Polychlorinated Dibenzofurans (PCDFs) and Polychlorinated Dibenzo-p-dioxins (PCDDs) from the Pyrolysis of Polychlorinated Diphenyl Ethers", Chemosphere, *9*, 351, 1980.

4.R. D. Grififn, "A New Theory of Dioxin Formation in Municipal Solid Waste Combustion", Chemosphere, *15*, 1987, 1986.

5.L. Stieglitz, G. Zwick, J. Beck, W. Roth, H. Vogg, "On the De-Novo Synthesis of PCDD/PCDF on Fly/Ash of Municipal Waste Incinerators", Chemosphere, *18*, 1219, 1989.

6.L. Stieglitz and H. Vogg, "On Formation Conditions of PCDD/PCDF in Fly Ash from Municipal Waste Incinerators" Chemosphere, 16, 1917, 1987.

7.J. R. Pedersen, "Isotopic Carbon Exchange Between a Simple Chloroalkane and Dioxins", Chemosphere, *18*, 2311, 1989.

8.Hileman, Bette, "Feed Supplement Contains Dioxins", Chem. & Engr. News, April 8, 2002. Kelp and copper sulfate heated together produced trace levels of toxic equivalents of dioxin at part per trillion levels as predicted by the de novo synthesis discoveries of independent investigators in 1986, such as H. Hagenmeir, Stieglitz, Nestrick and Lamparski.

THE BLUE RIBBON PANEL

"Believe one who has proved it, believe an expert", Virgil Horace,
70-19 B.C.

1979

At the press conference that released the "trace chemistries of fire" hypothesis, Dr. Robert Bumb announced that Dow would ask a blue ribbon scientific panel to study the analytical methodology used to obtain the dioxin data and evaluate its suitability and effectiveness. He said that Dr. Etcyl Blair, Vice President in Charge of Health and Environmental Sciences, would convene such a panel. Dr. Blair assigned me the task of finding a scientist "of world renown" to structure such a group. I was pleased to do this. Since the analytical methodology was new, it had not yet undergone peer review and was not yet recognized by other scientists as valid. A favorable report by the blue ribbon panel would reassure scientists and provide good public relations.

After extensive discussions with scientists with whom I had served on the EPA's Science Advisory Board's Committee on Environmental Measurements, I succeeded in persuading Professor Lockhart B. Rogers of the University of Georgia to head such a panel. Professor Rogers thus accepted the "opportunity to experiment with a new approach to the evaluation of science questions of public concern". In so doing, he would write his own criteria and rules.

Rogers would independently select members of the review panel, issue all press releases, and release the final report to the public. The panel would operate in an open manner, have access to all pertinent original data, have freedom to probe in depth all chemists and analysts involved in generating the data, and have access to the complete

protocol of the dioxin study. Each panel member would work approximately one week. Dow would pay expenses and service fees as
required by panel members.

Professor Rogers recruited the following:

Ursula M. Cowgill, Professor of Biology and Anthro-
pology, University of Pittsburgh Henry Freiser, Professor of Chemistry, University of Arizona Michael
L. Gross, Professor of Chemistry, University of Nebraska Dennis Schuetzle, Supervisor, Chem. & Physics Analysis Research, Ford Motor Co.

Members of the panel were supplied with the "trace chemistries
of fire" report and any requested related documents, before they came
to Midland.

Professor Ursula Cowgill was the first panel member to arrive in
Midland. She came on a Sunday, two days before the panel met, in
order to "see your facilities related to the dioxin issue and to walk the
banks of the Tittabawassee River". Her early arrival startled several
members of Dow's Dioxin Task Force, who suspected an ulterior motive
on her part. So, they met in the research conference room that Sunday afternoon and waited for me to bring Professor Cowgill from the
airport. She came loaded with samples of cores, which she had collected from borings in Israel many years earlier. (She thought the
analysis of these cores for dioxins and other chlorinated hydrocarbons
would shed light on the "trace chemistries of fire" hypothesis. The
cores had been taken from the determination of metals and had been
exposed to organic matter and thus probably contaminated. So they
were unsuitable for chlorinated hydrocarbon measurement.) These,
together with her baggage and her outdoor attire, made it immediately clear to the task force that she meant business and would do
whatever was necessary to determine the scientific truth. The entire
group was amazed. This seemed to be a new kind of professor! After
Professor Cowgill walked the river with John Gledhill, a Dow manager of Environmental Affairs, the rest of the panel arrived. The first
day they met privately with Terry Nestrick, Lester Lamparski, and
Lew Shadoff. The Dow scientists were asked to describe any new

developments to the analytical methodology since the report was written. This description should emphasize the weak points in the methods used in the "trace chemistries of fire" work. Observers representing the American Chemical Society and the Michigan Toxic Substance Control Board were allowed to attend these discussions. The Dow Dioxin Task Force was not allowed to attend, but we sat together in a room nearby. We were consulted only once the entire day.

Apparently the Dow scientists did a superb job describing the limitations of the methodology. By the end of the day members of the panel appeared to us to have the "blues". We took them to dinner at the "best" restaurant in the area. They ate in one room and we in another. Across the street, in another restaurant, analytical scientists ate together and were subject to be called by the panel.

Negative discussions proceeded through the entire dinner. By the end of the evening Dr. Bumb and others were certain that we were going to get a very negative report and they pressured me to somehow intercede. Although I was concerned by the tone of my peers, I was confident of the quality of the methods and the integrity of the members of the panel. "Sleep well!" I told Dr. Bumb, "Tomorrow is a new day!"

As I previously implied, I had excused myself from the deliberations of the panel. I had previously been accused by environmentalists of being a "company man", and doing and saying only what the company ordered. So I made myself available to answer any unanswered questions members of the panel may have, but made sure that I was not perceived as unduly influencing the judgement of the panel or controlling the remarks of the Dow scientists.

However, on the start of the second day's discussion, I appeared with the Dow scientists, asked if anything were needed from me and reminded them that I would be available in the same building. I also reminded them that on the previous day they had discussed what was wrong with the methodology and suggested that they spend the first hour discussing what was right with the methodology. They agreed and I left.

By the end of the second day the spirits of the panel members appeared to be considerably higher. We now waited for their report.

The final report was submitted to Dow on June 14, 1979. The
major conclusions of the panel were:

1. The analytical methods developed by Dow for dioxin
   detection and measurement are sound and valid.
2. Results above the detection limit appear reliable.
3. External analytical confirmation was advised.
4. More samples should be run in duplicate.
5. A clean room would be helpful for future work.

The increased confidence that the panel's study placed on the
data appeared to substantially reduce the number of sensational attacks on Dow scientists. Apparently the work was appropriately presented to the public. This occurred despite the effort by some to
place the credibility of the Roger's Scientific Panel in jeopardy. A
local newspaper article (Midland Daily News, March 14, 1979)
quoted an aquatic biologist with the Michigan Department of Natural Resources (DNR), alleging that "the chairman of the panel
(Lockhart B. Rogers), a professor of chemistry at the University of
Georgia, had done a great deal of consulting work for Dow. I'll let
you decide from there who's being impartial." The reporter interviewed Professor Rogers at noon of the first day of the panel's meeting in Midland and asked him for his reaction. He replied that he
had never consulted for Dow.

In a later report entitled, "Analytical Chemistry for Dioxins: Operation of Dow's Special Panel", submitted for publication to the
journal *Analytical Chemistry*, Professor Rogers gives the following account. "Before leaving Midland, the writer called the reporter to answer any further questions and learned that she was in difficulty with
her employer because of inaccurate reporting. It seems that, when
she had contacted the biologist following her noon-time interview,
he said she had misquoted him; what he had said earlier was that the
writer had analyzed many samples for Dow. When she asked him
where he got that information, he could not remember but suggested
that it was in some widely-read source, such as "Science" or "Chemical and Engineering News". In response, the writer promptly denied

Prof. Lockhart Rogers of the University of Georgia appreciated the "Wizards" expertise.

having analyzed any sample for Dow or knowing of any report to that effect."

"When people at Dow challenged the biologist about the newspaper report, he again indicated that he had been misquoted and that he had made unsuccessful attempts to contact the reporter to correct the original report. Later the writer wrote the biologist for possible clarification, but the biologist could not remember anything that might have led the reporter to make those statements."

Meanwhile Dow had written to the aquatic biologist demanding an explanation of the "calumnious, irresponsible and unprofessional statements" which "represent an unjustified attack on the integrity and business interests of Dow, Professor Rogers and the scientific panel members". In response, the biologist "would inform you that the statements attributed to me by the Midland press are inaccurate and misleading."

I was very unhappy about all of this. I had previously been interviewed and correctly quoted by the reporter. I had also been attacked unfairly by the biologist. Nevertheless, the reporter disappeared from the local news columns while the biologist continued his agenda with the state agency.

References

1. W. B. Crummett and G. L. Kochanny, Jr., "Academic consultants help industry", Chemtech, April 1983, pp 211-213. W. B. Crummett, "Multidimensional interfacial phenomena: professors, professionals, and the public", TrAC, 6, V (1987).

2. Starr Eby, "Dow disputed by DNR", Midland Daily News, March 14, 1979.

3. "DNR backs off on bias charges", Midland Daily News, April 12, 1979.

4. Public Forum, "Rohrer says he was misquoted", May 29, 1979.

5. Donna Sanks, "TCDD testing state-of-the-art", Midland Daily News, June 25, 1979.

6. Jeann Linsley, "Dow's dioxin analysis study lauded, but TCDD source still eludes", The Bay City Times, June 26, 1979.

7. Bob Solt, "Scientists praise Dow's testing procedures", Saginaw News, June 26, 1979.

8.Donna Sanks, "Study backs Dow finding, Bumb says", Midland Daily News, July 5, 1979.

9.L. B. Rogers, "Analytical Chemistry of Dioxins: Operation of Dow's Special Panel", Unpublished.

10.U. M. Cowgill, H. Freiser, M. L. Gross, L. B. Rogers, and D. Schnetzle, "Evaluation of the Analytical Chemical Procedures of The Dow Chemical Company for Determination of Polychlorinated Dibenzo-p-dioxins", June 1979.

ON BEING POLITICALLY CORRECT

"There is something inside of all of us that yearns not for reason
but for mystery . . . not for penetrating clear thought but for the
whisperings of the irrational . . . not for rationally founded
influence but for magic",
Karl Jaspers, "Reason and anti-reason in Our Time".

1984

Given its druthers, a research task force will tell its creator what the creator wants to hear. I had personally noted this as a member of various advisory groups and task forces for chemical companies, government agencies, academic institutions, and service clubs. It usually falls to those making measurements to ascertain that the report is valid. Often a web is spun around these members and their opinion is subdued or ignored. The result is a report, which flows in the same direction as the prevailing thought and is thus "politically correct".

Often the charter given the task force will automatically preclude their reporting anything other than a desired result. I experienced this as chairman of the Michigan Race Horse Industry Task Force. We were asked to develop a plan to assure that races would be run by horses that were free of drugs.

We developed a 26-step comprehensive plan, which would ascertain that the horses were indeed free of drugs. However, the plan was never instituted because the media was not looking for a solution. It apparently wanted a horror story. With about a dozen microphones in my face and several TV cameras grinding away, I spoke of the plan. But because I did not show the drugs that we had confiscated, the media did not report the study, and drugs at the racetrack become a non-issue. The Michigan State Department of Agriculture, the sponsor of the task force, was very pleased. Unfortunately, our 26 steps to a drug-free racetrack were forgotten.

The "Trace Chemistries of Fire" Task Force was entirely different. Our report pleased no one – not Dow management, not the U. S. Environmental Protection Agency, not the Michigan Department of Natural Resources, not environmental advocate groups, not the scientific community, not the automobile industry, and not the tobacco industry. But worst of all it seemed to enrage the news media.

To have been, "politically correct", in the view of most Dow management, we would have found no dioxins. But once we found some, it would have been "politically correct" to blame it on a point source, which could have been easily eliminated. "Trace chemistries of fire" meant that dioxins could not be eliminated. In fact, the threat that incineration would be under attack was not in Dow's interest since incineration was more valuable to Dow than all the chlorophenol plants. Having discovered "trace chemistries of fire", however, it would have been "politically correct" to share the data with the EPA administrators, the government, and the scientific community before releasing it to the public. Woe were we!

To have been "politically correct" in the view of the EPA, we would have had to report that the dioxins in the Tittabawassee River fish came from Dow's 2,4,5-T production plant. This was their supposition. Incineration, on the other hand, had been endorsed and recommended by the agency as a means of waste disposal. The Michigan Department of Natural Resources had the same "politically correct" position.

Since the goal of the environmental alarmists was to ban the use of sprays, being "politically correct" with them would also concur with the EPA.

With the news media, the "politically correct" position is impossible to ascertain. Probably any finding at that time would have been attacked.

For a scientist, being "politically correct" is not an option.

DECADE VII
DATA
INTERPRETATION

ASSESSING RISK

"Say first of God or man below
What can we reason but by what we know", Alexander Pope.
"It is not the facts which divide men but the interpretation of the
facts", Aristotle.

The ability to predict the acute toxicity of chemicals to man from animal studies is well established. Thus one can feel totally confident that a single exposure to a chemical at a concentration below a specific level will cause no immediate harm or even discomfort. Generally the same acute effects are observed for all animal species and at the same concentration levels.

For 2378-TCDD, the situation is not so simple. The concentration at which one half the animals die (LD50) varies with the animal as shown in the table below:

LETHALITY OF 2378-TCDD

Species	Sex	LD50, ug/Kg.*
Guinea Pig	Male	2.1
Rat	Male	22.0
Rat	Female	45.0
Rabbit	Mixed	115.
Hamster	Male	5100.

*ug/Kg means 0.000001 grams of 2378-TCDD per 1000 grams (2.2) pounds of body weight.

One thus has the problem of deciding which of these doses to extrapolate to humans to determine the level of concern for a single exposure to dioxins. Often the guinea pig data are used. This produces a level of concern at a concentration level which is much lower than has been the human experience, especially that of workers with chemicals known to contain dioxin.

The process for measuring the chronic toxicity by use of animals is much more rigorous and requires more than 20 difficult steps. Each of these steps, from designing a protocol to final examination of tissue slides, must be carefully defined and controlled to minimize uncertainty. The uncertainties cause deep and vigorous discussions among toxicologists whose points of view differ. On at least ten difference occasions, I have listened as such informal heated debates proceeded for a day or more with no consensus. As a result, the most fearful participant's position usually prevails. Such debates are typical of a science in its early stages of development.

In spite of these difficulties, more comprehensive work has been done on the toxicity of 2,3,7,8-TCDD than any other chemical compound. Detailed information is available on pharmacology, toxicology, pharmacokenitics, metabolic rates, and chemical-biological mechanisms. Some information is available on tissue doses. Considerable data have been taken related to ecotoxicity. Extensive studies in the field of epidemiology have documented the human experience. Even so, an EPA Science Advisory Board Committee of experts could not find a scientific reason to choose between a model which calculated a "cancer risk-specific dioxin estimate for 2,3,7,8-TCDD" of 0.006 picogram per kilogram per day and one which calculated 0.1 picogram per kilogram per day. Regardless of which number is ultimately chosen, the United States will be at odds with other nations of the world who have calculated 1 to 10 picograms per kilogram per day. Could it be that in the United States dioxin is really something other than a molecule? Or are EPA scientists more arrogant?

At the Fifth International Symposium in Bayreuth, Germany, September 1985, scientists from great nations, one by one, reviewed the scientific evidence, stated their assumptions (including reasonable safety factors), revealed their statistical models, and reported their conclusions. Those from Canada, Germany, Switzerland and Italy all agreed that the maximum allowable intake for 2,3,7,8-TCDD or its equivalent was 1-10 pg/Kg/day. Then an EPA scientist reported.

The 0.0064 pg/Kg/day "virtually safe dose", which he advanced, was shocking. The apparent arrogance of the American stunned us all! Discussions, subsequent to this announcement, have become largely political.

The numbers are now being called, "Virtually Safe Dose" (VSD) by agencies of the United States, and "Allowable Daily Intake (ADI) by other nations. A simplified table is:

Nation	Agency	VSD or ADI (pg/Kg/day)
United States	EPA	0.0064
United States	FDA	0.0572
United States	CDC	0.0276
Canada	OME	10.
Germany	West	1-10.
Netherlands	DINH	4.
Switzerland	SIT	10.
United Kingdom	UKDE	1-10.

(Note that, in the USA, risk assessors have the supernatural power to predict the "VSD" to three significant figures starting with a questionable 1 significant figure.)

All of the calculations shown in the table use the dose/response data reported by Kociba, et al. This 2-year feeding study is probably the most successful and comprehensive controlled study ever done on a single molecule. All the very best protocol was used, including incorporating the dioxin into the animal feed. Further, the malignant tumors were counted by an EPA consultant as well as Kociba himself with close agreement. At the lowest dose level (0.001 ug TCDD/Kg/Day) the liver and the fat of the rats each contained 540 ppt TCDD at the end of the study. This is significant since humans are not known to accumulate this much.

As one of an international group of scientists who organized conferences and as a member of various task forces and advisory groups, I sat with and listened to the world's toxicologists discussing ways to extrapolate this extraordinarily well-taken data to human risk. In 1988 in Ottawa, Canada, I understated my disappointment. "How best to extrapolate the animal toxicological data to humans has been a matter of considerable discussion and debate. As of now no consensus seems to exist." Today, 2001, no consensus exists!

Bruce Ames, renowned biochemist, has put the situation in perspective by comparing 2378-TCDD to ethyl alcohol. He concludes that the EPA reference dose of 6 femtograms of TCDD per kilogram per day in teratogenic potential is equivalent to one beer over a period of 8,000 years and in carcinogenic potential to one beer every 345 years. He concluded that the great concern over TCDD at levels in the range of the reference dose, "seems unreasonable". Even so, the National Toxicology Program has just listed 2378-TCDD as a "known human carcinogen".

More recently, the debate on how we deal with risks has become more vigorous. Some, including John Palen, appear to believe that the non-expert public has a more sophisticated view of risk than "experts" do. "Whether the risk is voluntary or imposed; whether it has catastrophic potential; whether it is equitably distributed; and whether it poses a threat to future generations", according to these people, are not addressed appropriately by scientists. Some say that the difference between "experts" and lay people are differences in values and perception of the facts.

Social scientists tell us that the public perceives involuntary exposure as riskier than voluntary, synthetic materials as riskier than natural materials, risks highlighted by the media of much more concern than those that are not. All of these add to the scary perception of dioxin. The scientist knows that the public's worry about a given risk is usually out of proportion to the real risk. Unfortunately the most credible and knowledgeable scientists are heard only by other scientists at scientific conferences. One such conference in which scientists took issue with the public view of risk occurred at the 207th American Chemical Society National Meeting in San Diego, California. There Dr. Alvan R. Reinstein, a physician at Yale University School of Medicine, concluded, "We must insist on rigorous science, no advocacy." This conclusion was reached after he had demonstrated the flaws riddling most epidemiological research. Once a scientist believes a substance is noxious, he is sure to find an effect at barely detectable levels. Most epidemiological studies and risk assessments depend on statistical evaluations and are subject to "bias, selective data analysis, phantom correlations, and logical error". The significance of such work can only be understood if it is compared with similar data on other substances. Chlorine-containing molecules compared with oxygen-containing molecules, for example, might add a dimension to human thought. A survey of nearly 1300 health professionals in the fields of epidemiology, toxicology, medicine and other health sciences urged that "risk assessment be made as scientifically objective as possible, more accurate and understandable, and consistency be established for more effective allocation of limited funds". Clearly, those on the front lines believe that objective science is the way to understand health concerns.

References

1.H. H. Hillman, "Certainty and Uncertainty in Biochemical Techniques", Ann Arbor Science Publishers, Inc., 1972.

2.R. J. Kociba, D. G. Keyes, J. E. Beyer, R. M. Carren, C. E. Wade, I. I. Dittenber, A. P. Kalnina, L. E. Freauson, C. N. Park, S. D. Barnard, R. A. Hummel, and C. G. Humiston, "Results of a Two-year Chronic Toxicity Study of 2,3,7,8-Tetrachlorodibenzo-p-dioxin in Rats", Toxicology and Applied Pharacology, 46, 279, 1978.

3.M. Richardson, "Toxic Hazard Assessment of Chemicals", The Royal Society of Chemistry, Burlington House, London, 1986.

4.Bruce N. Ames, "Cancer Scare Over Trivia", L. A. Times, May 15, 1986.

5.J. A. Young, "The problems of Assessing Toxicity", Chem. & Engr. News, July 13, 1987.

6.H. P. Shu, D. J. Paustenbach and F. J. Murray, "A Critical Evaluation of the Use of Mutagenesis, Carcinogenesis, and Tumor Promotion Data in a Cancer Risk Assessment of 2,3,7,8-Tetrachlorodizenzo-p-dioxin", Reg. Tox. And Pharm., 7, 57 (1987).

7.J. A. Moore, "Risk Assessment Reappraisals", Science, 240, 1125, 1988.

8.Bruce N. Ames, "Chemicals, cancers, causalities, and cautions", Chemtech, 591,

(1989).

9.R. E. Keenan, D. J. Paustenbach, R. J. Wenning, and A. H. Parsons, "Pathlogical Reevaluations of the kociba et al (1978) Bioassay of 2,3,7,8-TCDD. Implications for Risk Assessment", J. Toxicol. Environ. Health, *34*, 279 (1991).

10.R. J. Kociba, "Rodent Bioassays for Assessing Chronic Toxicity and Carcinogenic Potential of TCDD", Banbury Report 35: Biological Basis for Risk Assessment of dioxins and Related Compounds (1981).

11.W. E. Harris, "Analyses, Risks, and Authoritative Misinformation", Anal. Chem., *64*, 665A, Jul. 1, 1992.

12.M. M. Feeley and D. L. Grant, "Approach to Risk Assessment of PCDD's and PCDF's in Canada", Reg. Toxic. And Pharm., *18*, 248 (1993).

13.John Palen, "New risk assessment takes in more than just science", Midland Daily News, January 16, 1994.

14.Eric Larsen, "Mainstream media not doing the job on risk assessment", Midland Daily News, February 6, 1994.

15.D. Coleman, Abstract.

REASSESSMENT

*"There are in fact two things, science and opinion; the former
begets knowledge, the latter ignorance",
Hippocrates, c. 460-377 BC.
"We seem to be reaching the point of having to regulate the
regulators."
H. A. Laitinen, 1979.*

Nowadays (2001), when I discuss the regulation of dioxin effluents with industrial scientists, they dismiss the lack of consistency
with a shrug and say, "The science isn't there!" I respectfully disagree.
In my view the science is there at reasonable concentration levels. At
lower concentrations society is well served by common sense.

In 1995 the EPA reported that the largest source of dioxin releases was incineration:

Source	Percent
Incineration	31
Backyard trash burning	19
Landfill fires	19
Non-ferrous metal smelting	11
Power/Energy generation	4
Forest, brush, straw fires	4
Municipal sludge	4
Cement kilns	3
Others	5

In 2000 EPA announced that backyard trash burning had replaced incineration as the number one source of dioxin. This source
is subject to state restrictions only. So it is important for EPA (in

their view) to find a way to regulate backyard trash burning. They already have a possibility underway.

To assure that dioxin's "billion dollar industry" continues to flourish, it is important to have dioxin classified as a "known human carcinogen". Then the Delaney Clause would pertain and the EPA would have power to regulate all businesses and individuals beyond reason, assuring that all those in the "billion dollar industry" would thrive.

Analytical scientists and instrument manufacturers could speed up their inexorable drive to the ultimate limit – single molecule detection. Environmentalists would collect more donations. Trial lawyers would have more cases. Analysts would have more samples. Toxicologists would be attempting to discover more mechanisms to support the pronouncement. Epidemiologists would be free to monitor more employees. Journalists and authors could write many more scary stories, which would be politically, if not scientifically, correct.

Stricter regulation of dioxin would result, but it would not make the environment more safe. Dioxin is already regulated at levels too low to be monitored effectively. However, such regulations would make people even more fearful.

The EPA started the drive to achieve "known human carcinogen" status for dioxin in the 1980's and about every five years makes another reassessment. The review is a complicated process. The EPA Science Advisory Board first approves the draft study for review by a subcommittee called the Dioxin Reassessment Review Subcommittee (DRRS). The DRRS holds a public meeting where all special interests have an input, not necessarily scientific. A report is written and thoroughly criticized. The important question is whether dioxin is a "known human carcinogen". Two-thirds of the DRRS have recently (2001) said it is not. Based on the preponderance of the evidence, I would agree. Predictably, the EPA is not satisfied with this outcome and is exploring ways to change the meaning of the report.

One way to accomplish this is to disqualify scientists who have any working relationship with industry from membership on EPA's Science Advisory Board. This would eliminate much of the technical expertise and assure that the politically correct decision would be made. Dr. Donald G. Barnes, himself an authority on the science of dioxin, is trying to be fair and make the selection of members on technical know-how rather than politics but special interests make it very difficult. I suggest that a possible solution might be to compose the board of scientists from other countries only.

EPA's very first administrator, William D. Ruckelshaus, urged the agency to stop advocacy. Apparently no progress has been made. Veiled advocacy is still evident in the interpretation of data.

References

1. Callahan, J. M., "First U. S. Environmental Chief Urges Agency to Stop Advocacy", Michigan Living/Motor News, December 1979.

2. Harris, W. E., "Analyses, Risks, and authoritative Misinformation", Anal. Chem. 64, 665A, July 1, 1992.

3. Bellon, B. M., "Science Panel Meets on Dioxin", Midland Daily News, May 15, 2001, p. 1.

4. Heilpein, J., "Dioxin Decision", Midland Daily News, May 16, 2001, p. 1.

5. Hileman, B., "Reassessing Dioxin", Chem. And Engr. News, June 11, 2001, p. 25.

6. Hileman, B., "Dioxins Risk Assessment", Chem. And Engr. News, June 11, 2001, p. 8.

7. Hogue, C., "Conflict-of-Interest Debate Intensifies", Chem. And Engr. News, July 30, 2001, p. 38.

8. Gough, Michael, "Dioxins" Perceptions, Estimates, and Measures", "Phantom Risk", K. R. Foster, D. E. Bernstein and P. W. Huber, The MIT Press, pp 249-247, 1993.

INTEGRITY REGAINED

"In silence man can most readily preserve his integrity", Meister Eckhart, c. 1260 – c. 1327.

October 2-5, 1979

I made a horrendous mistake when I agreed to participate in an environmental conference at Lake George, New York. I misunderstood the invitation from Professor Lenore Cleseri, a fellow member of the Environmental Measurements Advisory Committee of the EPA Science Advisor Board. I thought I was to be a member of a panel to answer questions from the attendees. So I made no special preparation.

The setting was perfect for a conference. The weather was beautiful and the participants very friendly. Those in attendance were mostly college and university professors in the fields of journalism, philosophy, ethics, geography, engineering, and biology. Some chemists, including a few from industry, were also present. The different viewpoints of these various authorities provided stimulating and informative discussions and I became very relaxed, for once, in a conference on environmental matters.

But then I was shocked to note that I was listed on the program to give a talk. What a surprise! I was totally unprepared having no slides or data with me. What should I do? Talks on dioxin or "trace chemistries of fire" were much too involved for a 20-minute talk. However, I had just been involved in a Dow briefing describing an emergency response system designed to provide "instant" technical information and on-site support to local authorities who were suddenly managing leaking tank cars or pipe lines. So I made a second mistake. I reasoned that this group should know about this and decided to speak extemporaneously about it.

I told them why spills and leaks of hazardous chemicals were much more of a threat than traces in the soil, rivers and air. Such chemicals should not be shipped if it could be avoided. If shipped, the chemical should be in containers specially designed to prevent leaks even when damaged. I described these in some detail and concluded that parts per billion levels in lakes and rivers were of little concern compared to tank car spills.

My talk was exceptionally ill received. There were no questions – no discussion. People turned away from me. Persons who had been especially friendly the previous evening now kept their distance. Scientists from industry were nowhere to be found. Only Lockhart B. "Buck" Rogers remained friendly and I thought perhaps that was only because he had no choice. We had been assigned a room together. At dinner we were largely isolated from the rest of the attendees. Never had I been more devastated or unsure of myself! "Buck" and I debated my options far into the night. Should I disappear from the conference? We decided that I should not. There remained the questions of whether I should apologize for diverting the conference theme. I was not sure as I, with legs shaking, walked with "Buck" into the restaurant to breakfast.

A large number of attendees were in the lobby. They included those that I felt had snubbed me the evening before. Amazingly they all rushed to shake my hand and congratulate me on a fine talk. Totally confused we inquired what was going on. They pointed to the front page of a newspaper. A town about fifty miles away had been evacuated the previous night because a tank car of hydrogen chloride had derailed and spewed gas over the town. The writer of the article stressed the same points I had made shortly before this incident. Amazingly my integrity was made whole by the newspaper. Now viewed as ethical, we were surrounded by journalists, philosophers, moralists, and biologists who wanted to know more about environmental affairs in industry – especially "trace chemistries of fire" and "dioxin". "Buck" and I marveled at this strange coincidence and its impact on the emotions of rational thinkers. But I, for one, thrived on it.

THE UNEARTHED

"We can detect the dioxins more sensitively, using chemical –
analytic methods, than any other compounds on earth",
William W. Lowrance, 1985.

1979 – 1999

Immediately after the "trace chemistries of fire" hypothesis was announced, some eminent American scientists suggested ways to test the theory. The most popular suggestion was to search for chlorinated hydrocarbons and dioxins in ancient samples. Some scientists were prepared to provide samples for the study. For example, an oceanographer sent me condensed sea breezes from the South Pacific. An anthropologist sent drill core samples of highly carboneous layers of soil from under an ancient lake bed corresponding to ages of 5-, 12-, and 35 thousand years. Another delivered a 2,500-year old petrified fish. An archeologist sent ashes from the Biblical cedars of Lebanon. Unfortunately these samples were originally taken for other purposes and had not been adequately protected from contamination by chlorinated compounds. So analysis of these samples would be meaningless.

Other scientists suggested sampling ice from the center of a glacier. Still others wanted analysis of ash taken from the mouth of a volcano. Unfortunately we were unable to fund expeditions to obtain such samples. But even if proper sampling were possible, the analytical methodology was inadequate. We believed that detection limits at least as low as a part per quadrillion would be required because we had no knowledge of the degradation rates of these materials over long periods of time.

So, sea breezes, lakebed cores, ancient fish, glacier ice and volcanic ash were not analyzed. The contributing scientists were disappointed but understanding.

Some other scientists suggested the burning of fossil fuel in the presence of chlorine and inorganic chlorides. And yet others proposed the analysis of fish taken from rivers having various types and intensities of human activity on their banks. These two approaches appearing realistic and were undertaken by many scientists in several different countries. The results are discussed under the heading, "Testing the Hypothesis" elsewhere in this book.

One of the most diligent and effective scientists searching for meaningful samples was a friend and colleague, Dr. Ronald Kagel. On December 10, 1980, he received a very surprising but highly interesting letter from Lawrence A. Ernst of the Milwaukee Sewage District. He reported that he had a "sample of *Milorganite* produced in 1933 and displayed at the Chicago World's Fair!" It was "in a one pint Mason jar sealed with a zinc glass-lined cap". "We have no reason to believe this sample has been opened since January 9, 1934." A sample from that jar was procured by Dr. Kagel and analyzed by the Wizards together with samples from modern batches of (1981 and 1982) *Milorganite*. The results are summarized below.

Observed Dioxin Concentrations In Milwaukee Milorganite

	Concentration, ppt		
	1933	1981	1982
2,3,7,8-TCDD	2.	11.	16.
Total TCDDs	34.	138.	222.
Total HCDDs	1,500.	1,400.	1,400.
Total H$_7$CDDs	9,700.	9,400.	7,800.
OCDD	59,000.	50,000.	60,000.

These amazing results showed that dioxins were present in particulate matter before the introduction of chlorinated aromatic compounds into commerce. In fact, the hexa-, hepta-, and octa-chlorinated dipenzo-p-dioxins were at the same concentrations in 1933 as they were in 1981. This added considerable strength to the "trace chemistries by fire" contention that dioxins are ubiquitous.

In the meantime the Wizards had examined standard dust samples provided by the National Bureau of Standards. These samples were collected in St. Louis, Missouri, and Washington, D. C., analyzed and certified as standards for the determination of trace metals. The results are summarized in the following table:

	Concentration					
	TCDD		TCDF	HCDD	H_zCDD	OCDD
Airborn Dust	2378-	Other	2378-			
St. Louis, Missouri (SMR#1648)	47.	488.	380.	1.	32.	149.
Washington, D.C. (SMR#1649)	6.7	170.	102.	4.	37.	173.

These surprising results led to the Wizard's recommendation that analytical chemists analyze these standards to prove that their methodology was suitable for the analysis of soil and dust. The National Bureau of Standards approved the idea but could not certify the samples because certification required analysis by the Bureau and such analysis required more effort than could be justified. So they now supply standards along with the Dow results.

Others confirmed the Wizard's data and the results provided more evidence that dioxins are ubiquitous. Since the concentration is higher in St. Louis, a manufacturing site, than in Washington, the results add to the data which show dioxins are always higher in samples taken from sites where more energy is being expended.

Scientists, including myself, continue to desire data on ancient samples. Late in 1984, I learned that the Environmental Contaminants Sections of the Environmental Protection Service, Edmonton, Alberta, Canada planned to look for dioxins in flesh from the man who froze during the ill-fated Franklin Expedition to find the Northwest Passage in 1850 and remained frozen. On January 3, 1985[2], I wrote to Chief, Mary Anne Sharpe, as follows: "Judging from quantities of chlorinated dioxins found in soil sealed in glass in 1877 (Rappe 1985) and other environmental data, one might expect to find somewhere between 1 and 100 parts per quadrillion of various chlorinated dioxins in the adipose tissue of adult humans in 1850. These are very low levels and the results can be seriously affected by either contamination or by the use of insufficiently sensitive and specific analytical methodology. For these reasons the results, no matter what their magnitude, are likely to become highly adversarial. To protect yourself against unreasonable attack, it may be wise to subject your protocols to peer review before undertaking the analysis. Better still would be to have the samples analyzed by a number of laboratories who agree on quality assurance protocols."

While we were very cautious about generating data on which we could not reasonably depend, others appeared to ignore any limits or restrictions and grabbed samples without protocol whenever and wherever the opportunity arose. Thus a New York physician[3] rushed to collect adipose tissue samples from 400-year old frozen Eskimo cadavers. As could have been predicted, the laboratory to which the samples were referred had problems because they were working at the limit of their method's detection. Reanalysis showed the presence of low part per trillion levels of hepta—and octachlorinated dibenzo-p-dioxins. Even with all this uncertainty the authors conclude that the main contemporary source of PCDD/F's is anthropogenic. To reach such a conclusion the tacit assumption was made that PCDD/F's are stable in frozen human adipose tissue. That is, any degradation must have a half-life greater than 100 years. As of this writing I know of no way to test that assumption. To accomplish such a task more sensitive analytical methodology is required, together with long-range experiments covering at least 50 years, before such a determination can be made.

The Eskimo data could reasonably be interpreted in another way but it would not have been "politically correct" to do so. "Politically correct", had already been defined by scientists at Indiana University[4]. They measured PCDD/F's in dated sediment core samples from the bottom of Siskiwit Lake and reported that their concentration had increased greatly since 1940 corresponding to the production of synthetic chlorinated compounds by industry. To reach this conclusion they assumed that the PCDD/F's were stable in lake sediment over the time period of 1910 to 1980. Since the best information available on their stability in soil in the United States shows that 2,3,7,8-TCDD degrades with half-life of 10-12 years[5], the assumption is highly speculative. Another speculation would be to assume a half-life of 10 years in lake sediment and constant fall-out from the air. This would also account for the perceived increase after 1940, producing a curve almost identical to the one reported by the authors.

A somewhat more definitive study[6] was made on mummies of Chilean Indians. This time a qualitative study of the stability of lipids was made. Based on this study the PCDD/F's were thought to be stable. The tacit assumption is that molecules at or near infinite dilution do not degrade faster than the matrix in which they are found. Such an assumption is not consistent with scientific experience. If a half-life for PCDD/F's in dry muscle is assumed to be 100 years or even 400 years, they would not be found by modern analytical techniques. The fact that 2,3,7,8-TCDF and 2,3,5,7,8-PCDF were reported to be found, is remarkable.

A much more comprehensive and well-designed study was made on soil and herbage from an agricultural experiment plot of ground in southeast England[7]. Samples of top soil, taken from the same untreated experimental plot in 1846, 1856, 1893, 1914, 1944, 1956, 1980, and 1986, were dried and stored in glass bottles sealed with a cork, at ambient temperature in the dark and analyzed for chlorinated dioxin and furans at the same time after 1986. The results were astonishing. The total PCDD/F's increased from 31 to 92 parts per trillion from 1856 to 1986. These results were plotted with no further treatment with the tacit assumption that no degradation had occurred. The resulting curve suggested to the authors that atmospheric fall-out of PCDD/F's began to increase about 1900. This may be true but it is equally likely that it not be. If one assumes a constant fallout and a half-life of 60 years, a similar curve is obtained. Much more work needs to be done on the stability of PCDD/F's in various matrices if we really want to know how much is "natural". This will require better controlled and longer term experiments together with more sensitive and reproducible analytical methodology.

Recently a well-designed study was made of sea breezes in the North Atlantic Ocean[11]. Sampling was done from towers in Bermuda and Barbados. Enough PCDDs and PCDFs were found to calculate a deposition rate over the oceans to be 1 ton per year compared to 12 tons per year over the earth's land mass.

Perhaps more importantly, a recent study has shown that "sediment associated 2,3,7,8-substituted dioxin residues in general, and 2,3,7,8-TCDD in particular are in a state of flux, as they are produced from peri-dechlorination of octaCDD and further laterally dechlorinated to 2-MCDD[12]. This supports the idea of half-lives being necessary to properly interpret environmental dioxin data.

All of the studies mentioned here are very interesting. Some are beautiful works. Some are not. They suggest many things but prove nothing. After studying them, I am persuaded that it is possible to discern how much PCDD/Fs are formed from natural sources. I am also convinced, however, that I was wrong in 1985 to discourage Chief Sharpe in her desire to analyze flesh from the frozen man of the Franklin Expedition. I now think readily detectable amounts of PCDD/Fs would have been found, consistent with the study of English soil. It is no longer clear that people are responsible for most of the dioxin in the environment. Still those[8,9,10] who would ban chlorine chemistry cite these studies as proving "dominance of anthropogenic sources". Until scientists understand the degradation rates and treat them appropriately, such conclusions are unwarranted.

The recent discovery of dioxins in ball clay[14] in Mississippi (at depths corresponding to millions of years) with a totally different congener distribution than previously observed, suggest a geologic mechanism. Similar observations were made on ball clay in Kentucky and kaolinitic clay in Germany. Several different mechanisms of formation and degradation to explain these observations are plausible, but none can be proved.

As I write this, a new exciting report says that chlorinated organic molecules are formed when plants decay. A new highly technical technique called near-edge x-ray absorption fine structure spectroscopy is able to distinguish between chloride ions and aromatic and aliphatic organo chlorine without sample preparation. This finding may account for these materials being present in some "unpolluted environments."

So I hereby humbly apologize to all those scientists who appeared to interpret the "trace chemistries of fire" hypothesis as meaning that most dioxins occur naturally. They may be more nearly correct than I was. After a thorough review of these works, I am persuaded that the source of dioxins in the environment is of more natural origin than I would have predicted from the "trace chemistries of fire" hypothesis in 1978.

If dioxins do not degrade in soils and sediments, as tacitly assumed by these investigators, fear mongers will have reason to create panic during floods or other earth movement, such as *Newsweek* magazine did in 1986 in reporting the flood in Midland[13], Michigan. Their report showed more concern about the spread of dioxins than the welfare of people made homeless.

References

1. W. B. Crummett, "Environmental Chlorinated Dioxins from Combustion – The Trace Chemistries of Fire Hypothesis", "Chlorine Compounds and Related Compounds", O. Hutzinger, R. W. Frei, E. Meriam, and F. Pocchiari, editors, Pergamon Press, 1982, pp 253-13.

2. W. B. Crummett, Letter to Mary Anne Sharpe, Chief of Environmental Contaminants Section of the Environmental Protection Service, Edmonton, Alberta, Canada, January 3, 1985.

3. Schecter, A. Dekin, N. C. A. Weerasinghe, S. Arghestani, and M. L. Gross, "Sources of Dioxins in the Environment: A study of PCDDs and PCDFs in Ancient Frozen Eskimo Tissue", Chemosphere, *17*, 627-631, 1988.

4. J. M. Czuczya and R. A. Hites, "Airborne Dioxins and Dibenzofurans: Sources and Fate", Environ. Sci. Technol., *20*, 195-200, 1986.

5. A. L. Young, "Long-term Studies on the Persistence and Movement of TCDD in a Natural Ecosystem", "Human and Environmental Risks of Chlorinated Diox-

ins and Related Compounds", R. E. Tucker, A. L. Young, and A. P. Gray, editors, Plenum Press, 1983, pp 173-190.

6. W. V. Ligon, S. B. Dorn, R. J. May and M. J. Allson, "Chlordibenzofuran and Chlorodibenzo-p-dioxin Levels In Chilean Mummies Dated to about 2800 years before the Present", Environ. Sci. Technol., 23, pp 1286-1290, 1989.

7. L. O. Kjeller, K. C. Jones, A. E. Johnston, and C. Rappe, "Increases in the Polychlorinated Dibenzo-p-dioxin and –furan Content of Soils and Vegetation since the 1840s", Environ. Sci. Technol., 25, 1619-1927, 1991.

8. T. G. Spiro, and V. M. Thomas, "Sources of Dioxin", Letter to the Editor, Science, October 21, 1994.

9. B. Dudley and P. Cosner, ibid, October 21, 1994.

10. T. J. Spiro and V. M. Thomas, "Organochlorine Compounds", Letter to the Editor, C & E News, February 13, 1995.

11. J. I. Baker, and R. A. Hites, "Polychlorinated Dibenzo-p-dioxins and Dibenzofurans in the Remote North Atlantic Marine Atmosphere", Environ. Sci. Technol., 33, 14-20, 1999.

12. I. D. Albrecht, A. L. Barkovskii, and P. Adriaens, "Production and Dechlorination of 2,3,7,8-tetrachloro-dibenzen-p-dioxin in Historically-Contaminated Estuarine Sediments", Environ. Sci. Technol., 33, 737, 1999.

13. "Dioxin Scare", Newsweek, September 29, 1986.

14. J. Ferrario, C. Byren, and D. Cleverly, "Summary of Evidence for the Possible Natural Formations of Dioxins in Mined Clay Products", Environ. Sci. and Tech. (Accepted for Publication), 2000.

15. Myneni, S. C. B., "Formulations of Stable Chlorinated Hydrocarbons in Weathering Plant Material", Science 295, 1039 (2002).

NATIONAL COUNCIL OF CHURCHES

"There is always an easy solution to every human problem —
neat, plausible, and wrong", H. L. Menchen.

4 February 1980

I once took my mother to visit an old friend who lived in the
Shenandoah Mountains. To reach her, we had to take a gravel road up
a hollow. These roads were unmarked and several appeared to be the
right one. To identify the correct hollow, I stopped at a country store
to ask directions. From the sounds inside the store, a party was in
progress. I heard a vigorous conversation, punctuated by sounds suggesting the shuffling of chairs and the thud of tobacco spit hitting
the spittoon.

On my stepping inside, all activity stopped. I could hear a fly
buzzing in the room. Nine pair of hostile eyes stared me down. They
seemed to say, "What the hell are you doing here, stranger? How dare
you interrupt our relaxation around the pot-bellied wood stove?" I
felt a chill although the pot-bellied wood stove was radiating an intense heat and the season was early autumn. I stood there immobile,
petrified with uncertainty. Finally, regaining some small amount of
composure, I asked where the Neselrodts lived. All eyes relaxed, became friendly, and all nine owners of eyes competed to help me find
my way.

Not so, the eyes of the people at the National Council of Churches.
They were icy cold on our arrival and remained so throughout the
visit. Certain groups associated with the National Council of Churches
had purchased stock in The Dow Chemical Company. On learning

about Agent Orange, they had questioned the ethics of the company and felt an obligation to impose their moral values on the company. To this end, they proposed to embarrass the company at its annual stockholders meeting by proposing that a special advisory committee be appointed to determine if the company should discontinue the manufacture and sale of chlorinated phenols and derivatives. Taken by itself, this proposal appeared reasonable. However, many credible scientific groups had already considered this question and the company was following their conclusions. The company believed that the folks at the National Council of Churches were unaware of these studies and asked for an opportunity to present those findings. This request was granted and a few of us were sent to New York to review the information.

On arrival, we were greeted stiffly and ushered into a conference room. Immediately, the cause for the cold, stiff reception was apparent. One of our chief accusers, a notable physician, sat on the opposite side of the table. He reluctantly shook hands, sat down, and was joined by council members. There we sat – Dow personnel on one side of the table and on the other, our defamer with church representatives. On cue, he launched a bitter, abusive criticism and condemnation of Dow research and policy, blaming us for many of the ills of society. It was a diatribe we had heard from him before. No wonder trusting church people were misinformed, I thought.

Finishing his harangue on toxicology, our vilifier started a tirade about analytical chemistry. He announced with great pomp that dioxin had been found in human mother's milk and this surely meant an epidemic of cancer. He was referring to work at Harvard University in which Patrick O'Keefe had obtained signals from a few human milk samples. These signals were slightly higher than background, but were the same as the limit of detection. These signals may or may not have been due to dioxin. No one could say for sure. I quickly praised him for keeping abreast of scientific investigations, expressed concern for the mothers in Oregon, who were needlessly scared by this, and asked him to comment on the most recent EPA report. He looked blank. "Have you not seen the EPA report", I asked. "No, I haven't", he said. "Well, here it is." I handed him a copy and pointed out that no positive results were found in any of the 103 samples examined. He stared at it incredulously. Then he quickly said, "What am I doing discussing such ridiculous matters with stupid people like you. I have more important things to do. Good day!" And he quickly got up and left the room.

I was elated! He had just demonstrated that he had no use for data, no matter how valid, if it did not support his agenda. If he were not chief protagonist, he'd take his prejudices home. Now the good folks at the National Council of Churches would surely see that they had been misled. I could not have been more wrong. I could feel the anger and bitterness in the room. It was as though I had insulted a saint. Several turned their backs when later I extended my hand to say goodbye. To think that a molecule could cause virtuous people to behave this way defies reason. Could dioxin really be an evil spirit?

Reference

1. "Eight Church Groups File Dow Action," Paul Rau, Midland Daily News, March 19, 1980.

UNCERTAINTY

*"If we do measure it there is associated with the measurement a
specified range of uncertainty", F. D. Rossini.*
*"The central problem of our age is how to act decisively in the
absence of certainty", Bertrand Russell.*

"Well! Here I am again – back in uncertainty!" I opened my talk
on "Analytical Methodology for the Determination of PCDDs and
PCDFs in Products and Environmental Samples: An Overview and
Critique", in the Second International Symposium on Chlorinated
Dioxins and Related Compounds, October 25, 1981, in Arlington,
Virginia. I expected the audience to howl with laughter. Instead there
was dead silence. Obviously, they did not appreciate the truth or the
humor in such a statement. Perhaps they did not trust my motivation since I was from industry. Perhaps they thought I was trying to
pass off the horrors of dioxin as a joke. It was a sobering moment!

It was the most appropriate remark! It was the least appropriate
remark! Here I was standing to present the best analytical methodology on trace analysis the world had yet seen. Still I was very aware of
its uncertainty. Apparently most did not see the big picture. Or perhaps they felt that if they admitted uncertainty in their data, others
would not believe it to be credible and it would not be accepted in a
court of law. At that moment all I could do was proceed to present
my information.

For almost 20 years I had preached to Dow people that hard,
firm decisions should be based on credible analytical data. The data
generated by our laboratory was the final answer, close to absolute,
because its integrity was assured. I was speaking of the usual measurements at concentration levels of 0.01 to 100.0%. At parts per

billion and lower levels, the uncertainty becomes much greater and it became necessary to emphasize it.

Generally analytical chemists strive to attain 95% confidence in their data. Sometimes chemists claim 99% confidence. Such reports are suspect and often result in the data being very critically reviewed by others. At 1 part per trillion a variation in results of 150% is likely.

Twelve years after my talk, the concept of uncertainty is recognized and taught. In major cities in the United States a course is being offered entitled, "Measurement Uncertainty" by the Measurement Technology Company of Newbury Park, California. This is very encouraging. The science of measurement requires an understanding and measurement of uncertainty. The practice of good sound science requires taking uncertainty into account. Thus regulations specifying precise number limits are based more on politics then science and regulators should so state.

TO FIND OR NOT TO FIND

"Whatever is only almost true is quite false, and among the most dangerous of errors, because being so near the truth, it is the more likely to lead astray", Henry Ward Beecher.

1981

In 1981, fish from Michigan rivers remote from The Dow Chemical Company, were analyzed by the laboratories of two different highly reputable public institutions. The results appeared to be very different as reported by the news media.

Michigan State University found TCDD in the fish at levels as high as 586 parts per trillion. These results produced a furor on national television news and headlines on the front page of some newspapers. These results were contrary to what the government and the media wanted to believe; namely that the "trace chemistries of fire" hypothesis was wrong and that the only TCDD in fish came from the herbicide production plants of the chemical companies, especially Dow. So, much was done to discredit and embarrass the professor who was responsible for the study. Picking up on this, some extremists even threatened him. After all, they appeared to think it was the "politically correct" thing to do.

In the middle of the furor, the U. S. Food & Drug Administration announced to the media that they had analyzed fish from the same Michigan rivers and had found no TCDD at detection levels of 25 parts per trillion. The announcement appeared to cause great consternation in the media, which now found it difficult to be "politically correct". Generally the "politically correct" position was to believe data from universities because they were "independent" and had no "vested interest", but accepting the Michigan State results

239

was equivalent to endorsing Dow's "trace chemistries of fire" hypothesis which was "politically incorrect". On the other hand, it was "politically correct" to attack government regulatory bodies such as the FDA, but in this case believing their data was also "politically correct", since their data threw doubt on the "trace chemistries of fire" hypothesis.

I secretly resolved to avoid contact with reporters on this matter since neither study had yet been reported in the scientific literature. However, my resolution was short lived. In Washington, D. C., for a scientific conference, I was approached by an attendee who identified herself as a reporter for the Washington Post. "Who is right? Michigan State or FDA?", she asked. Spontaneously and with great conviction, I replied, "Both are right!" She was incredulous and insisted that they can't both be right. "You will have to go on television and explain this to the American people", she said, turning away and leaving abruptly. My scientist friends were pretty shaken over this exchange as well. "What in the world are you saying?" they asked. "One laboratory was qualitatively right but quantitatively wrong. The other was qualitatively wrong but quantitatively right." "We see", they said politely excusing themselves.

Since I have not been invited to appear on television to explain my logic, I will now attempt to do so. Two years earlier, our laboratory (the Wizards) had analyzed 17 fish from some of the same Michigan rivers, remote from Midland. TCDD had been found in about 9 of the samples. Most of the concentrations were in the range of 5-11 ppt, well below 25 ppt, FDA's limit of detection. But a few (about 5%) were above that level with the highest being 104 ppt. Because the media furor was so intense we did not yet have Dow approval to publish this data.

Because I had complete faith in the Dow data, I felt that the Michigan State laboratory was correct in reporting the presence of TCDD but the concentrations appeared to be too high. I also recognized that it would be possible to take samples of fish, all of which had concentrations below 25 ppt, and so the analysis by the FDA laboratory could be right but the number of samples too small. Since I did not have access to either data set, I could not comment further.

Later I included the Dow data in a table entitled, "Search for TCDD in Fish", presented at the Third International Symposium on Dioxins held at Salzburg, Austria, 12-14 October, 1982. No sensational reaction to this data occurred. Meanwhile the Michigan State University professor (Dr. Matt Zabik) was still being harassed and vilified by the media and environmental groups, a judgement totally unjustified.

Still later I learned that a team of experts from the U. S. Environmental Protection Agency, headed by Bob Harless, had examined the Michigan State work in minute detail and found everything in order in the methodology and the treatment of data. Later still, I was informed that the FDA laboratory did indeed find a few fish that had levels of TCDD in the order of 100 ppt but discarded the results, claiming the samples had been allowed to stand too long before analysis. The likelihood that the TCDD content would change over a period of several weeks is very small unless the laboratory was highly contaminated. There are valid reasons for excluding a data point from a set of data, but even so, all the results should be reported and explanations given. Since, to my knowledge, the FDA has never reported their results to the scientific community, we'll never know for sure.

Industry has no choice. It must report all data as precisely and as openly as possible. To be sure the data will not be misinterpreted, great care must be used in presenting it. That is why, especially when using newly developed analytical methodology, the time honored title, "Apparent Concentration" is useful. That is what I used in 1982.

There is something uncanny in the ability of inexperienced chemists to find what they are looking for at the limit of detection. They may inadvertently analyze the wrong samples, mistake impurities for analyte, unintentionally contaminate the sample, or fail to make the appropriate separations. Having once believed they have found what they are looking for, it is almost impossible to convince them otherwise. There is something inherently rewarding in finding what one is searching for and so it is easy to report false positive results. Most experienced analytical scientists, however, will make every effort to find the analyte and will find satisfaction in their work whether they find it or not.

It is now possible to tailor analytical methods to assure the finding of a signal, which can be identified as the sought analyte, whether it is present or not. Likewise methods can be devised to guarantee that the analyte will not be found. Persons needing information should be aware of this and only accept data that has been challenged in the scientific arena.

The practice of analytical science is a sacred trust. We must subordinate our own agendas in order to advance science and thereby contribute mightily to society. In this endeavor we desperately need the help of the media who will also need to subordinate theirs.

Generally disparate data sets taken by different schools of science are of little concern, for science has its own time-tested system for resolving differences. Talks at scientific conferences, together with publication in peer-reviewed journals, allows the scientific community to make rational judgements without introducing non-scientific agendas into the debate. But this process cannot be rushed.

Reference

1.W. B. Crummett, "Status of Analytical Systems for the Determination of PCDD's and PCDF's", Chemosphere, *12*, 429-446, 1983.

GREAT EXTRAPOLATIONS

"What! Will the line stretch out to the crack of doom?",
Williams Shakesphere, MacBeth IV, i. 117.

In making measurements everyone agrees that accurate calibration is needed for the results to be reliable. We also agree that the measurement should be made within the calibration range. So distances are measured based on the wavelength of light. Volumes are measured based on carefully calibrated containers. Weight is measured by balancing with carefully calibrated weights. All measurements refer to primary standards developed and controlled by national governments.

Since analytical chemists produce calibration curves, which are used to calculate concentrations of analyte, extrapolation of data is seldom done and is, therefore, not a problem. However, as concentrations are measured at very low levels (parts per trillion), reliable calibration curves are exceedingly difficult to produce. For example, to measure 1 part per trillion of TCDD in a given matrix with a high degree of confidence, data should be taken on known concentrations of TCDD from 0.01 to 100 parts per trillion in the same matrix. The difficulty in preparing such matrices is so very great that it is seldom done. Instead, the calibration is made at part per billion or part per million levels and extrapolated to zero concentration. This results in uncertainty since we cannot be sure the curve will extend in a straight line or curve up or down.

Although such extrapolations are serious and require careful consideration before they are used, they are probably the least troublesome of extrapolations. Much more serious and totally unscientific are the investigations that find qualitative evidence for the presence of a pollutant at a particular site and connect it to a possible point source many miles away. An example of extreme extrapolation, is found in which a report which cited the chemical waste burner at Dow Chemical Company's complex in Midland, Michigan, as a source of "dioxin" found in sediment at the bottom of Siskiwit Lake. Siskiwit Lake is an inland lake on Isle Royale National Park in Lake Superior and is approximately 300 miles from Midland. Later the record was set straight at a scientific conference at Michigan State University. The author said her study shows that a "small part of the source could be natural combustion".

In another study, dioxins were found in fish in the Grand River near Lansing, about 80 miles from Midland. The newspaper accounts suggested that these dioxins were coming from Dow's Midland complex, even though the Grand River does not flow near Midland, nor do any of its tributaries.

Yet another example is the allegation that dioxins found on flood plains are from paper mills 65 to 100 miles upstream even though the dioxins found were very different from those normally found at paper mills.

Generally the environmental alarmists will say that almost any "perceived" undesirable human or ecological effect is "linked" to dioxin if analysis shows dioxin to be present at any level. This is a "humongous" extrapolation.

On the mountain farm where I grew up, chains were very important. Especially remembered are logging chains, harness chains, tire chains, dog leashes, and watch chains. The characteristics of the links determined the strength and use of the chain. Some logging chains were about half an inch thick and made of iron. These chains would not break even when two steam locomotives were linked and pulled against each other. At the other extreme were the decorative toy chains made by we kids of white pine needles. These chains were very fragile and great care was required in handling to keep the links from springing apart.

"Link" is one of the words used frequently in magazine and newspaper articles. Some molecule loosely called a "toxic substance" is linked to some human malady. The reader naturally tends to think of the links as a part of very strong chains. In reality, the "links" are more like the pine needle loops. In all such statements that I have investigated the "link" is based on either very preliminary, semi-scientific, or pseudo-scientific studies. Therefore, conclusions are not yet warranted. Using this concept any molecule, having shown a biological effect in any animal at very high dosage levels, can be "linked" through a long series of extrapolations and assumptions to that effect in humans, or any other species.

"Links" are usually much more political than scientific. The use of this term is very misleading.

In 1995, the EPA released yet another "risk assessment and risk characterization" of dioxin in a 2000 page draft report. This last in a series of reports dating back to the 1980's is the first to heighten concerns about the non-cancer effects of detectable exposure[1]. These concerns were immediately applauded by scientists associated with advocacy groups saying, "the risks are greater than previously thought"[2] and more regulation is needed. But other scientists[3,4] said the scientific evidence was insufficient "to support EPA's alarming conclusions", and there was no evidence that "dioxin exposure compromises immune function in humans". When scientists disagree to this extent, we can be sure that the data on which judgement is being made is not very firm and the "links" are fragile.

Scientists, like all humans, display their fallibility by sometimes assuming an extreme position on data interpretation. Those on one extreme require mountains of "perfect" data before they can draw a hypothesis. Those on the other extreme require only one data point to develop grandiose explanations. Generally scientists in industry, whose judgements may affect the health of coworkers and themselves, need much more "fool-proof" data to make recommendations than academic scientists who live remote from the area in question.

So it seems easy for some scientists to extrapolate from the mouse to humans, the test tube to the troposphere, and a point source to the universe. But other scientists are stuck with needing reliable analytical measurement in the mouse and human, the test tube and the troposphere, and the point source and the universe. At least scientists on the right need data taken as close as possible to the end of the extrapolation. Since budgets will never allow some to collect all the data they need, they can only doubt the pronouncements of others.

Readings

1. Jeff Johnson, "Dioxin Risk: Are We Sure Yet?", Environ. Sci. Technol., *29*, 24A, 1995.
2. R. P. Clapp, P. DeFur, E. Silbergeld, and P. Washburn, "EPA On The Right Track", Environ. Sci. Technol., 29, 29A, 1995.
3. R. S. Greenberg, J. Margolick, D. R. Mattison, P. Munson, A. B. Okey, D. L. Olive, A Poland, J. V. Rodricks, L. Rudenko, and T. Starr, "EPA Assessment Not Justified", Environ. Sci. Technol., *29*, 31A, 1995.
4. C. A. Bradfield, et al., (12 others), "EPA Dioxin Reassessment", Science, *266*, 1628, 1994.

TRACE CHEMISTRIES OF OTHER MATTERS

*"The infinitely great and the infintesimally small cannot be
exhaustively known or explored, or conjectured about", Liehtsa
5th or 4th Century B.C.*

After the "trace chemistries of fire" experience, I felt as though my
mind had been released from the energy barrier which had until then
prevented me from seeing the importance of innumerable types of molecules interacting at very low concentrations in all systems, living or
dead. I had, because of my limited experience and simplified education,
restricted my thinking too much to that which could be sensed by the
five senses, suppressing thought which in any way challenged the pre-
cepts learned in a formal school setting.

Some of us (Dow Dioxin Task Force) now saw a much larger picture.
We were certain that others would see it too. A new frontier would thus
be identified which would create challenges for analytical scientists far
greater than previously dreamed. "Trace chemistries of fire" would be
only one small area of interest. The trace chemistries of water, air, plant
sap, blood, digestive systems, human cells, sea breezes, forests, asphalt,
concrete, etc., might be others. Research on these systems we thought,
would proceed vigorously. Beginning with further studies on the "trace
chemistries of fire" we expected the new vision to extend to others –
especially the most important, physiological fluids, emphasizing human
blood.

The National Academy of Sciences, striving mightily for intellectual
clairvoyance and grasping for straws of academic enlightenment through
great extrapolation in order to be "politically correct", has squeezed out a
possible relationship between parts per trillion levels of TCDD and some

rare diseases found in a few Vietnam veterans, even though the diseases could have been caused by various other toxic exposures. If the scientific community trusts this conclusion, then research is urgently needed on the "trace chemistries" of human blood. As Bruce Ames has pointed out, natural carcinogens at part per million levels occur in most common foods. So these toxins get into the blood stream. What happens to these and the many other natural chemicals in the human body is not known. Research programs on this question are not easily found.

TRACE CARCINOGENS IN VEGETABLES AND FRUITS[2]

Food	Carcinogens
Cream of mushroom soup	hydrazines
Carrots	caratoxins, myriaticin, isoflavones, nitrate
Cherry tomatoes	hydrogen peroxide, nitrate, quercetin glycosides, tomatins
Radishes	glucosinolates, nitrate
Cranberry sauce	eugenol, furan derivatives
Lima beans	cyanogenetic glycosides
Broccoli spears	allyl isothiocyanate, glucosinolates, goitrin, nitrate
Baked potato	amylase inhibitors, arsenic, chaconine, isoflavones, nitrate, oxalic acid, solamine
Sweet potato	cyanogenetic glycosides, furan derivatives, nitrate
Pumpkin pie	myristicin, nitrate, safrole
Apple pie	acetaldehyde, isoflavones, phlorizin, quercetin, glycoside, safrole.

Since most of the natural molecules of concern are water soluble and are not stable for long periods of time, the measurement is much more difficult at trace levels than are the persistent molecules such as TCDD. Also, because of their high activity they may react quickly with tissue in the body forming metabolites at greatly reduced concentrations. The effect of these reactions on health is unknown. Perhaps more importantly, those remaining together with their metabolites may eventually bathe every cell in the body. So they may be potentially more dangerous than the persistent molecules, which are mostly trapped in the fat.

If we are truly fearful about the toxicity of molecules at concentration levels of parts per billion or lower, we must recognize the fact that each of us is built around the most efficient complex chemical digester in the world – our own digestive system. We fuel our digesters with products taken from other living organisms. These products contain traces of toxins in addition to nutrients. Therefore, we are all managers of molecules, illustrated by comic strip characters[4]. Our digestive systems, together with our immunological systems are so effective that the mutagens and carcinogens are handled quite well, but nothing 100% certain. However, if we are truly concerned about toxins in our diet at less than a part per billion, we should change our reasons for eating and drinking certain things.

If we define pollution as the presence of extraneous molecules in a system, then the human blood stream is the most highly polluted flowing system and human adipose tissue the most highly polluted stationary system. Analytical scientists note this by every separation system devised to isolate analytes of interest. The number of different molecules called interferences at a part per billion level often exceeds 10,000. Because these behave in many different ways they present an unmeasureable number of challenges to our immune systems.

Still, we need not fear the infinite nor the infinitesimal. We can measure and control neither. However, with reasonable effort, we can manage molecules down to concentrations of a part per billion or thereabouts. Below that we should have no fear because we have no control. We can pray for control but we put our faith in jeopardy by so doing. We can do research to learn how to manage below a part per billion but there is no urgency in such research and little reward for the researcher.

References

1. R. L. Renso and P. Recer, "Veterans Affairs Office Adds 2 Disorders to Agent Orange Help List", Midland Daily News, July 27, 1993.

2.David J. Hanson, "New Agent Orange Study Links Herbicide to Diseases", C & E News, August 23, 1993, p 15.

3.American Council on Science and Health, "Holiday Dinner Menu", 1991.

4.Guisewite, Cathy, "Cathy", Midland Daily News, October 21, 2001.

5.Bruce N. Ames and Lois Swirsky Gold, ibid., pp. 153-181, 1993.

DECADE VIII
PR TRIUMPHS

THE DOW DIOXIN INITIATIVE TEAM

"For we must look under every stone, lest an orator bite us",
Aristophanes (446-380 B.C.).

1982-1984

On June 1, 1983, I sat among the reporters at a Dow press conference at which Paul F. Oreffice, President and Chief Executive Officer of The Dow Chemical Company bowed to the forces of regulation and government interference in the management of private industry. With David Buzzelli, Chairman of the Dow Dioxin Team, and James Sanders, a Dow physician and member of the "Team", standing tall beside him and reinforcing his statements, Mr. Oreffice announced that Dow would spend $2.5 million on dioxin studies. These studies would include: a search for new dioxin sources inside the plant, using outside auditors to monitor the investigation; a $250,000 case control study by the state health department on a possible link between "dioxin" and soft tissue sarcoma in Midland County; a study by an unnamed independent national scientific organization on the question of whether trace quantities of dioxin in the environment pose a health hazard to humans; a $250,000 Dow grant to the University of Michigan to develop technology to remove dioxin from waste water discarded into rivers; and the purchase of additional analytical equipment and the development of more scientists skilled in the determination of dioxin in environmental samples.

Headlines which resulted from the announcement:
1 Jun 1983. Lorie Shane, Midland Daily News, "Dow to Spend $2.5 Million on Dioxin Studies".

1 Jun 1983.   Associated Press, Boston Globe, "Dow Says It Will Study Dioxin".

2 Jun 1983.   Ron Koehler, Bay City Times, "Hold Anti-Dow Talk Until After Tests: Reigle".

2 Jun 1983.   Booth News Service, Bay City Times, "Local Critics Hedge Praise for New Dow Dioxin Tests".

2 Jun 1983.   Des Moines Register, "Dow Believes it can Disprove Dioxin Danger".

2 Jun 1983.   David Everett, Detroit Free Press, "Dow Offers Cash for Dioxin Study".

2 Jun 1983.   Ann Cohen, The Detroit News, "Dow Starts Study on Dioxin Danger".

2 Jun 1983.   Lorie Shane, "State Checks to see if Dow Grant can be Accepted".

2 Jun 1983.   Minneapolis Star and Tribune, "Dow Plans Campaign to End Fears About Dioxin".

2 Jun 1983.   Iver Peterson, New York Times, "Dow announces Campaign on Dioxin".

2 Jun 1983.   Dennis Bell, Newsday, "Dow Funds $3-Million Dioxin Study".

2 Jun 1983.   Rocky Mountain News (Denver), "Dow Program Targets Dioxin Fears".

2 Jun 1983.   Susan Benkelman, Saginaw News, "Dow Outlines $3 Million Dioxin Study Program".

2 Jun 1983.   St. Paul Pioneer Press, "Dow to Test Dioxin in Anti-scare Effort".

2 Jun 1983.   San Francisco Chronicle, "Dow to Finance Big Study of Dioxin in Michigan".

2 Jun 1983.   Wall Street Journal, "Dow Chemical Slates $3 Million for Studies on Effects of Dioxin".

2 Jun 1983.   Ward Sinclair, The Washington Post, "Dow Planing $3 Million Program to Allay Fears Over Dioxin".

Within a week the wisdom of editors appeared in various publications:

3 Jun 1983.   Detroit Free Press, "DIOXIN: Dow goes to work

on a contamination problem and a credibility gap". "Dow
Chemical's offer of 2.5 million for dioxin studies is a welcome acknowledgement of the public concern over the
possible effects of the chemical, not to mention the damage
the dioxin controversy has done to Dow's image. The
studies, if they are carried out under conditions that assure
their integrity, should help remedy both problems."

3 Jun 1983. The Flint Journal, "Dow and Dioxin". "Dow
Chemical's offer to provide money for scientific studies to
determine what health risks, if any, may be involved in the
byproduct, dioxin, is a better response than its earlier
protestations that the toxic material has been with us ever
since man discovered fire."

5 Jun 1983. Saginaw News, "Dow's $3 million gamble". "The
Dow Chemical Co. commitment of $3 million for research
on dioxin is at once a responsible and risky act of corporate
citizenship. It's responsible because something has to be
done to get at the truth about dioxin, finally, independently,
and definitively. The action is risky for Dow because the
company insists that it will keep its corporate hands off the
research."

7 Jun 1983. Midland Daily News, From the Wall Street Journal,
"Dioxin worries not based on evidence." "The notion that
dioxin is a doomsday menace is based less on medical
evidence than on some kind of psychological phenomenon".

8 Jun 1983. Chemical Week, "Clearing smoke in Midland".
"The challenge for Dow is to come to terms with today's
sociopolitical environment. Endowed with a feisty corporate
culture, Dow has resisted vigorously – and sometimes
justifiably – what it considers unwarranted demands by
government. In doing so, though, it has projected too often
an insensitivity to public concerns. It appears that Dow has
finally recognized the problem. Its new dioxin program,
while based on science, is driven by public relations. The
scientific results may or may not be reassuring, as the
company expects, but the PR effort is surely on the right

track. Indeed, openness – coupled, perhaps, with just a touch of humility – might do wonders for the image of what, in truth, is and always was a very fine company."

Although I had been a member of the Dow Dioxin Initiative Team and had participated in all their deliberations, including detailed discussions of scientific data and approaches, I felt uneasy about the program presented by Mr. Oreffice and did not fully agree with any of the assessments made by the editorials. When Paul told the news media that, "We will literally leave no stone unturned", I knew that the analytical scientists associated with me had extraordinarily difficult tasks ahead. Furthermore, the program required other Dow scientists to work with agencies driven by public relations rather than sound science.

The search for new dioxin sources inside the plant would require at least a year using all the qualified Dow analytical scientists. The difficulty of doing this work while fulfilling the promise to develop more analytical scientists for dioxin analysis and installing a new unfamiliar mass spectrometer was monumental. And before these could be started I would have to attract a world renowned analytical scientist (with the highest integrity) from a major university to serve as mentor, advisor and monitor.

The $250,000 grant to the University of Michigan was insufficient for the university to produce much more than a literature search. I knew that many universities had already attempted to remove dioxins from wastewater by a variety of approaches, all of which proved futile. If the university discovered a new and useful approach, analytical data would be needed and the only known laboratory capable of such work was Dow's own Analytical Sciences Laboratory, thus placing more unreasonable demands on our operation.

The $250,000 grant for a case control study by the state health department was outside my area of expertise but I wondered how such a study could be useful when there was no agreement among Michigan physicians as to what constituted "soft tissue sarcoma".

I knew too, that the unidentified national prestigious scientific society, whether the American Medical Association or the National

Academy of Sciences, would require input from the analytical science community, chiefly Dow.

To be blunt, but honest, I was sure that anything scientifically useful resulting from this effort would have to come from Dow's Michigan Division with the analytical laboratory being the major contributor. However, I was confident that the work would be done according to the best science and the probability that it would show Dow to have a clean plant was high. This is why I voted with the other members of the Dow Dioxin Initiation Team to do this work when Chairman Dave Buzzelli said, "It is time to vote! How do we vote? Read my lips." Much later I was informed what this meant.

For me the major concern was to engage outside auditors to monitor the investigation. All Dow people agreed that was my job. That was as it should be for I had insisted that an internationally recognized analytical chemist be asked to act as an outside third party auditor and monitor. This seasoned and independent chemist's function would be to examine every aspect of the experimental protocol from the gathering of samples through the multi-step, complex analytical procedures, to the documentation, interpretation, and presentation of data; to observe in detail, the planning sessions of the Dioxin Initiative Team, including the dialog and debate; and to correct any gaps or oversights in the study.

To fully appreciate the magnitude of the project, the following facts would have to be considered by the monitor:

* The Dow Midland plant site covers 1500 acres
* The Midland plant powerhouse can burn 2000 tons of coal per day.
* The Midland plant operates four natural gas augmented burners to incinerate halogenated by-product waste streams. One of these, the waste incinerator, is by far the largest, typically burning 205 tons of solid and liquid trash and waste per day.
* The Midland plant site wastewater system handles 19,000,000 gallons of water per day flowing through the 1500-acre plant site.

*   The number of samples, which can be analyzed, is limited
by the difficulty of the analytical method.

The analytical methodology must be understood by the auditor/
advisor for him/her to appreciate the difficulty of the task and to
contribute in the design of the investigation. For soil, the methodology consisted of: (1) extraction, (2) a primary clean-up procedure
consisting of three liquid flash chromatography columns, each designed to remove a special type of interference, (3) reversed-phase
high performance liquid chromatography (HPLC), (4) silica – HPLC
refractionation of TCDDs and (5) high resolution gas chromatography – low resolution mass spectrometry. As a result of this complex
and rigorous effort, all the 22 tetra-, 10 hexa-, 2 hepta-, and
octachlorodipenzo-p-dioxin are completely separated from each other
and from all known interferences. Their concentration is measured.
2,3,7,8-Tetrachlorodibenzofuran is also separated and measured.

The data generated by this methodology is elegant but it is of
little value unless it has been taken according to a rigorous quality
assurance plan to evaluate the laboratory performance. Such a plan
generates its own data set which includes: (1) internal standards for-
tification and recovery, (2) reagent blanks, (3) duplicate analyses and
reference standards, (4) identification and quantitation criteria, (5)
instrument performance criteria, and (6) demonstration of isomer
specificity.

All of the plant conditions, the analytical methodology, and the
quality assurance plan would have to be understood in detail by the
auditor for him/her to function with credibility.

Furthermore the auditor would need to interact with the news
media with confidence but without appearing arrogant. We needed
an auditor who already had knowledge of most of the information I
have just outlined. Among the many possibilities one man stood out.
He was Professor Henry Freiser of the University of Arizona!

To my great honor, Professor Freiser agreed to take on the assignment. In accepting this challenge Professor Freiser was undertaking
to oversee the selection, analysis, and interpretation of a limited number of samples, including several different matrices from a very large

site. The sampling, analyses, and interpretations were highly complex. In this effort, Professor Freiser would work closely with twelve Dow scientists with advanced expertise in waste disposal, statistical analysis, analytical chemistry, manufacturing process chemistry,

Prof. Henry Freiser glows as he explains his contributions in the search for "dioxins".

chemical engineering, theoretical chemistry, and incineration.

During the course of the study, Dr. Freiser spent one or two days per month in Midland with members of the team. When the report was ready to assemble, it was clear that its size and complexity would probably prevent its being published in a standard referred scientific journal. Dr. Freiser suggested and Dow agreed, that he recruit and select a panel for peer review. For this purpose Dr. Freiser invited R. Graham Cooks of Purdue University, Otto Hutzinger of Bayreuth University, and Peter Jurs of Pennsylvania State University. They accepted after being assured of total access to the data and total freedom to comment on the final report.

The review panel met in Midland where the Dow Dioxin Initiative Team presented a draft report. The panel members then studied the information over a period of a month before meeting in Chicago to write a detailed review and critique to be used by Dow for report revision. This review and critique was published as part of the final report. The report of the panel included discussions of sampling protocol, analytical methodology evaluation and interpretation of results, finger printing, mass flow, and point sources. Strengths, weaknesses, scope, and limitations were defined and explained. Nevertheless, the panel found the conclusions of the initial report to be substantially correct. The conclusions were:

* Levels of 2,3,7,8-TCDD outside the Midland plant site were below 1 part per billion.
* Levels near the plant boundary were 10 to 75 times higher than levels in typical industrial areas of other cities due to historical Dow practices.
* More than a mile from the plant, Midland soil was typical of other industrialized cities in 2,3,7,8-TCDD content.
* Two small areas within the plant site had elevated levels of 2,3,7,8-TCDD.
* The waste incinerator stack is the only current significant point source of 2,3,7,8-TCDD on the plant site.
* About 0.6 gram per year of 2,3,7,8-TCDD is released to the Tittabawassee River in the treated wastewater.

These headlines followed release of the report:

2 Nov 1984.   Keith Naughton, Saginaw News, "Track out of Dow dioxin worries EPA".

3 Nov 1984.   Midland Daily News, "EPA – No Advice for Dow Workers".

3 Nov 1984.   David Everett, Detroit Free Press, "Cleanup Planned for 2 Dioxin Sites at Dow Chemical".

3 Nov 1984.   Dudley K. Pierson, Detroit News, "Dioxin peril draws focus to Midland".

5 Nov 1984.   Bay City Times, "Dioxin hot spots at Dow contained, officials tell press".

5 Nov 1984.   Julie Morrison, Midland Daily News, "Dow says dioxin no threat".

5 Nov 1984.   Keith Naughton, Saginaw News, "Dow: Incinerator caused air, soils, dioxin problems".

6 Nov 1984.   Casey Bukro, Chicago Tribune, "Dow plant dioxin contamination found".

6 Nov 1984.   Keith Naughton, Saginaw News, "Dow dioxin ad mission contradicts barbecue theory".

6 Nov 1984.   Wall Street Journal, "Dow Chemical releases report on Dioxin Levels At Its Midland Facility".

6 Nov 1984.   Philip Shabecoff, New York Times, "E.P.A. and Dow Agree on Sealing Off Contaminated Soil in Michigan".

6 Nov 1984.   Philip Shabecoff, New York Times, "Dow To Cover Soil Tainted by Dioxin".

6 Nov 1984.   St .Louis Post-Dispatch, "Chemical Firm: Dioxin pollution No Health Threat".

6 Nov 1984.   Dudley K. Pierson, Detroit News, "Dow calls dioxin levels safe".

The tone of the headlines was more realistic than earlier ones. Whether due to the perceived humility of Dow management or the credibility of the outside auditor, the change was welcome. Nevertheless, an uneasiness was felt among Dow scientists. Sound science seemed to be taking a back seat to public relations. Could Dow science be a loser? This uneasiness was reinforced by the motto displayed by some Dow managers: "Perception is Reality".

The Dow Dioxin Initiative seemed to calm the media and herein lies its biggest accomplishment. Scientifically, the effort supported the "trace chemistries of fire" hypothesis and demonstrated that Dow had not seriously contaminated the plant site by leaks and spills. Although the University of Michigan study did not produce a practical system for removing dioxin from water, studies by Dow scientists did.

Nevertheless, the risk to Dow employees and Midland residents did not change. Although some dioxin was contained by asphalt covers, the risk simply went from insignificant to insignificant. In other words the concentrations were so low that the risk could not be measured with scientific reliability.

Since the media responded favorably, Dow managers recognized, congratulated, promoted and awarded each other. In addition, all members of the Dioxin Initiative Team got two tee shirts and a paperweight. I still have mine.

When the Environmental Protection Agency collected soil samples in the City of Midland, Dow scientists went with them and collected duplicate samples to be analyzed by the Wizards. The samplers were a strange group. The regularly dressed, bare-headed, bare-armed, and bare-handed Dow people were in sharp contrast to the EPA employees in white cover-all space suits, helmets, and gloves. Amazingly home owners did not cry out or appear flustered by having this strange task force invade their lawns and dig around their down spouts.

On April 8, 1997, the headline in the Midland Daily News read, "Study shows lower dioxin levels at Dow, in Midland". The article claims the dioxin levels have declined but "the state and the company agree more testing needs to be done."

Three days later (April 11, 1997) the Midland Daily News advised that "drawing conclusions from results difficult" under the headline, "Dig this about dioxin testing". Eight schoolyards and eleven parks were tested. Of these, at least one of each was higher than the state's regulation level. However, samples taken from communities "with no industrial base" showed positive results higher than some in

Midland. Although the Department of Environmental Quality (DEQ) folks "were surprised to get some level of detection at all" in these samples. The results were consistent with the "trace chemistries of fire" hypothesis.

A public meeting was held and less than ten citizens attended. This prompted a scathing letter to the editor, "Daily News was irresponsible in handling the dioxin study issues". The editor responded, "Midland Daily News is an advocate of the public, not environmentalists or industry". Both articles were published under the headline, "Making sense of dioxin coverage". However, nothing said in this exchange addressed the question: Should there be concern from these results?

On November 21, 1997, under the headline "Dioxin exposure not linked to death", by Lisa F. Smith, in the Midland Daily news, assured us that a Dow study revealed that "workers potentially exposed to dioxin haven't died earlier, nor have they died from cancer more often than the general population".

In 2001, a panel studied the dioxin situation in Midland and Mayor R. D. Black concluded that, "There is no evidence of any health consequences in Midland, Michigan." However, the editor of the Midland Daily News wants to keep doing soil tests to "remain vigilant against the harmful effect of dioxin". (What harmful effects?)

Still the Dow effluent to the river is regulated at "levels below one part per quadrillion, a level unmeasurable by most analytical laboratories. And 90 parts per trillion is said to be the "trigger level" for soil in residential areas.

More References

1. R. J. Agin, V. A. Atiemo-Obeng, W. B. Crummett, K. L. Krusnel, L. L. Lamparski, T. J. Nestrick, C. N. Park, J. M. Rio, L.A. Robbins, S. W. Tobey, D. I. Townsend, and L. B. Westover, "Point Sources and Environmental Levels of 2,3,7,8-TCDD on the Midland Plant Site of The Dow Chemical Company and in the City of Midland, Michigan". Dow Chemical Company, Midland, MI, November 5, 1984.

2. W. B. Crummett, "Multi-dimensional interfacial phenomena: professors, profes-

sionals, and the public", TrAC, 6, V, 1987.

3. W. B. Crummett, F. J. Amore, D. H. Freeman, R. A. Libby, H. A. Laitenen, W. F. Phillips, M. M. Reddy, and J. K. Traylor, "Guidelines for Data Acquisition and Data quality Evaluation in Environmental Chemistry", Anal. Chem., 52, 2242, 1980.

4. L. H. Keith, W. B. Crummett, D. Deegan, Jr., R. A. Libby, J. K. Taylor and G. Wentler, "Principles of Environmental Analysis", Anal. Chem., 55, 2210, 1983.

5. "Panel says dioxin levels no risk in Midland", Midland Daily News, October 14, 2001.

6. Editorial, Midland Daily News, October 23, 2001.

7. "Looking for DIOXIN", Midland Daily News, December 2, 2001.

SCIENTIFIC EFFORT
DISALLOWED

"I don't make jokes. I just watch the government and report the facts", Will Rogers.

March 1983

As chairman of the Analytical Division of the American Chemical Society, I had worked very hard to establish an "Excellence in Teaching Award" for professors of Analytical Chemistry. Finally, all plans were in place and I was scheduled to present the first award to Isaac M. Kolthoff at the society's annual meeting in Seattle, Washington, on Tuesday, March 22, 1983. I was very enthusiastic and excited because Professor Kolthof was generally recognized as the foremost scientist in the field of analytical chemistry. Furthermore, he had been the advisor for the Ph.D. degree of my own graduate studies mentor, Professor William MacNevin at Ohio State University, and thus my professional grandfather. Moreover, many of the best known analytical chemists of the nation and perhaps the world would be there.

About two months before the award ceremony, I was invited to give a talk on assuring the quality and integrity of analytical data at a joint meeting of analytical chemists of the United States Army, Navy, Air Force, and Marines. The Air Force captain who extended the invitation explained that the military had held these conferences for many years, but non-military persons were not permitted to attend. I was the first civilian scientist ever invited. Unfortunately, I was scheduled to speak on Tuesday at about the same time as the award presentation in Seattle, so I respectfully declined but felt greatly honored. My Air

Force champion, however, consulted the program committee and they decided to convene on, Sunday, March 20, to allow me to speak and get to Seattle and do my work there. I was thrilled and accepted immediately.

I felt I had reached the zenith of my career. If I could get the U. S. military to use the ACS guidelines generated by a committee I had chaired, other government functions would surely follow their lead and society would benefit greatly from the generation of better data. It was surely going to be the most exhilarating week of my professional life.

Then I heard that a U. S. congressman intended to hold hearings on dioxin in Washington, D.C., at some unspecified time, possibly the week of March 20, when I had such wonderful plans. Since I would probably have to be in Washington, if a hearing were held, I informed Dow's Director of Research, Dr. David P. Sheetz, of the situation. He advised me to give the talk in San Antonio and fly to Seattle. If I were needed in Washington, he'd have a Dow plane pick me up in Seattle. We agreed that I would need no special preparation for the congressional hearing.

On March 19, I arrived in San Antonio, was met by the air force captain who drove me to a motel. He then left but would pick me up in about 2 hours for dinner. Disrobed, I was about to step into the shower when the telephone rang. The ensuing conversation went something like this: "You are needed in Midland immediately", the Dow manager said. "Catch the next plane to Chicago. A Dow plane will meet you there. If you can't get to Chicago by nine o'clock tomorrow morning, the Dow plane will pick you up in San Antonio. Call me back and tell me what plans you can make".

I protested vigorously. "But, Paul", I said. "I am here in San Antonio at the invitation of the United States Air Force to address a joint meeting of analytical chemists of the army, navy, air force and marines. It is the first time ever that a civilian scientist has been invited to this meeting. The military has special concerns about assuring the credibility and integrity of data. I have personally worked fifteen years to introduce rigorous standards into the general practice of analytical chemistry. If the military as a unit adopts these principles and puts them into practice, it will have a profound positive effect in the quality of analytical data in the United States".

"I agree with you", Paul said. "But this is a matter of survival. You are to appear before a congressional committee chaired by Congressman Schuyer on Tuesday."

"Fine!" I said. "I will give the talk tomorrow, interact with the military scientists, and catch a plane to Chicago tomorrow evening."

"Impossible!" Paul retorted. "Return at once! Bob Bumb is coming from Europe!"

"Have Dave Sheetz call me and tell me I should come!"

"Dave Sheetz is out of town and unavailable for comment."

"Then have Ron Yocum call me!" Ron Yucum, to whom I reported, was the Director of Research of the Midland Division. My caller was a technical director and had no authority to give me orders.

Puzzled that my immediate supervisor was being by-passed, I yielded. It was one of the most regrettable things I ever did!

I apologized to my would-be host and gave him my slides together with my talk. He practiced giving it and did well enough to substitute for me on Sunday. I flew to Chicago early the next morning and was met by a Dow plane, which took me to Midland immediately. Unbelievably, there was no one there to meet me. My wife eventually got me home where I waited an entire day before being allowed to see the testimony already prepared and approved. I was especially chagrined because the military had scheduled me on Sunday, outside their regular meeting time, so that I could make my later commitments in Seattle. But there was nothing I could do!

So on Tuesday, six of us Dow people sat in a hearing room on Capitol Hill somewhat overawed by the attention of the media. Three Dow managers presented the testimony. Each was supported by the presence of a Dow scientist seated immediately behind them. My role was to support Bob Bumb on testimony related to incineration and combustion. During the presentation of the testimony, congressmen and their aides paid no attention but came and went in an apparent haphazard manner, conversed with each other, laughed at inappropriate times, waved to other persons, and in general displayed contempt for those giving testimony. After some questions, vaguely related to the testimony, we were subjected to vilifying harangues by each congressman in turn. However, it was obvious that the most paranoid, non-factual, vulgar, obnoxious tirade was reserved for the chairman. The horrifying thought that he was probably buying votes by this behavior was most nauseating.

I somehow survived sitting there hopefully expressionless, not showing my disgust for fear of a "contempt of congress" citation. A feeling close to contempt crept over me. Not only had I given up a rare opportunity to influence the U. S. military with the possibility of saving lives, but I had also missed a once-in-a-lifetime opportunity to be part of a scene at that moment being enacted in Seattle, Washington, to honor a great chemist.

I kept telling myself, "If you can keep your head when all about you are losing theirs and blaming it on you . . . ." I felt the peak of my career had been chopped off by men of political stature gone mad over a molecule. Or could it be something more powerful? Worse still, I was angry with myself for returning from San Antonio to Midland. I paid a high price for being honest. In the hearing I spoke nary a word. At least I can be thankful for that.

The most amazing thing in this whole episode was that the military had been understanding while I took orders from Dow management the only time in my career. Had I sold my soul to Dow management? If so, dioxin must be more than just a molecule.

I'm still not sure why I was called back so precipitously. Paul Dean was not the problem. He has approved this writing.

References

1. W. B. Crummett, F. A. Amore, D. H. Freeman, R. Libby, H. A. Laitinen, W. F. Phillips, M. M. Reddy, and J. K. Taylor, "Guidelines for Data Acquisitions and Data Quality Evaluation in Environmental Chemistry", Anal. Chem., *52*, 2242, 1980.

A BILLION DOLLAR BUSINESS

"We promise according to our hopes and perform according to our
fears",
La Rochefoncaid, Mezims.

24 January 1984

At a meeting of scientists at an environmental standards workshop on chlorinated materials in incinerator emissions at McLean,
Virginia, an EPA attendee emphasized that "chlorinated dioxins are
now a billion dollar per year business". It is not difficult to believe
that this statement is true. Evidence continues to mount. For example, in October 1987, the North Atlantic Treaty Organization
(NATO) Committee on the Challenges of Modern Society published
a list of 96 Dioxin Analytical Centers. If each center has only one
high-resolution mass spectrometer, that alone would represent a capital
investment of about 50 million dollars. I recently received an advertising pamphlet from a major manufacturer of mass spectrometers
accompanied by a letter dated February 17, 1993, which opened,
"Wouldn't it be nice to walk into your laboratory this morning and
find that all your analytical needs for high-resolution analysis of halogenated dibenzodioxins and dibenzofurans have been met? Wouldn't
it be nice to have the first high-resolution magnetic sector mass spectrometer which is truly dedicated for halogenated dibenzodioxin and
dibenzofuran analysis? Such a system is finally here!"

The development of such a system would not have been possible
without the perceived need on the part of many analytical laboratories for high-resolution mass spectrometry. This is not surprising,
however, for after about 1972, many commercial analytical laboratories tried to justify the purchase of high-resolution mass spectrom-

eters simply because they "were required for dioxin analysis". That society will spend billions chasing chlorinated dioxins at the part per trillion and part per quadrillion levels while neglecting other social ills, speaks to its insatiable thirst for the superstitions as opposed to the scientific.

Why is society still functioning on a philosophy which many scholars thought died in the Dark Ages? This question must be pondered by journalists, philosophers, psychologists, historians, scientists, politicians, and religious leaders alike. Each of these will put its own unique perspective on the question. The true answer, however, will not be a sum of the inputs, but those levels on which there can be general agreement.

On November 29, 1993, I received a brochure extolling the merits of "The Dioxin-Prep Sample Clean-Up System". In fifty years of practicing analytical chemistry, I have seen only one or two other systems named or dedicated for a family of molecules.

Now, 1995, the expenditure of $20,000 to $40,000 for an automated clean-up system; $200,000 to $1,000,000 for a gas chromatography-mass spectrometer system; $75,000 to $200,000 for a clean laboratory room; and about $400,000 per year for personnel, equips one to become a part of this "billion dollar industry". But one can also expect a start-up time of at least a year before useful data will be obtained. Even then, data of the highest integrity will probably not be obtained.

The thought that the great pioneering work of Dow's Analytical Sciences Laboratory, of which I was a part and am justly proud, contributed mightily to the building of the "billion dollar industry" saddens and sickens me. When I joined Dow, I was prepared to work diligently to create a large industry dedicated to producing goods to brighten the lives of people. The dioxin industry has done little but promote fear, and I am trapped in the irony of history. Recently (1997), EPA has proposed that dioxins be included on the Agency's Toxic Release Inventory. If approved, its principal effect will be to assure the continued growth of "the billion dollar industry".

During my 18-year tenure as technical manager of Dow's Analytical Sciences Laboratory, the laboratory invested about $1,750,000 in mass spectrometers, about $250,000 in liquid chromatographs and related equipment, and about $50,000 for special laboratory housing to measure dioxin. Personnel costs for the "trace chemistries of fire" study amounted to about $550,000. Total analytical chemistry costs over this period amounted to about $10,000,000. These costs do not include the cost of toxicological studies, epidemiological studies, statistical analyses, legal costs, consultation fees, product improvement, enhanced water treatment, incinerator improvement, and remedial action. Since I was not directly involved in the latter nine activities, I will not attempt to estimate the cost of each of these, but a conservative estimate of the total might be more than $100,000,000.

Since 1988 the laboratory founded by Nestrick and Lamparski has had its sample load increase four fold. This indicates that the "billion dollar" industry is flourishing.

Reference

1. "Dioxins proposed for toxics inventory", C & E News, May 12, 1997.

PERCEPTION

"In science, I think, we see but a shadow of our instruments",
Graham Cooks, April 11, 1984.

11 April 1984

Early man's awareness of the natural world was gained only through impressions provided by his sense organs. These measurements were qualitative only and consisted of hearing, seeing, smelling, tasting and touching.

As instruments were invented, new measurements and observations could be made. These resulted in new insights. The understanding thus gained depended not only on the observation, but also on an understanding of the capability and limitations of the instrument itself. This has lead to the interesting conclusion quoted above as said in a symposium in which Professor Cooks was presented the American Chemical Society Analytical Division Award in Chemical Instrumentation. Since new instruments are constantly being invented, the "shadows" being seen are constantly changing. Perception is thus constantly changing. Even when the instrument remains the same, the "shadows" change as the scientist learns how to make the instrument perform better.

Analytical chemists learn to control or brush away the shadows in order to make precise and reliable measurements. This enables them to observe the behavior of molecules and he/she can report their activities to the world. In this respect, they are like a newspaper reporter. The chemist reports on events which result from the behavior of molecules while the reporter reports on events which take place because of the behavior of human beings. Like the chemist, the reporter must be careful to tell the difference between the "shadows"

Dow Research Scientists, produced irrefutable data, and sang special tunes.

of his/her instruments", such as an unreliable source, and the truth.

So it is surprising to see that journalists, editors, historians, philosophers, and others read much more into the scientific discussion than the scientists intend to convey. As an example, an editorial in the Midland Daily News in August 1985, was headlined, "Dioxin concerns show democracy is working". "Recent breakthroughs in methods for destroying dioxin," it said, "are a tribute to public awareness as much as they are to science".

CHEMOPHOBIA

*"At this point in the last 1970's and early 1980's, the
mysterious TCDD had assumed the properties of black magic –
pervasive, unpredictable, and malevolent – in the minds of
many people. Like such magic, to them TCDD must be the result
of ill-intentional witches or warlocks and these could only be the
chemists and chemical companies who produced the noxious
material", Lawrence B. Hobson, M.D., Ph.D., 1985.*

20 September 1985

Mitwitz Castle, Bavaria, West Germany, playground of Herr
Doktor Professor Otto Hutzinger, was glorious on this rare fall day
when both nature and man were in tune, it seemed. The lunch in the
courtyard was delightful and a walk around the moat and through
the garden was idyllic. We spoke of how romantic it must have been
to have lived in such a place at the time the castle was at its greatest
prosperity. But then, it attempting to mentally return to that time,
we realized we could not totally visualize it for so many things were
now missing from the old castle. Most difficult to reconstruct were
the odors – those from the moat into which raw sewage had drained;
those from the butchering of animals in the courtyard every day (a
necessity since there was no refrigeration); the putrid smell of meat
infected with maggots; the smell of manure of all types, especially
horse manure; the wet feathered smell of plucked fowl; and worst of
all, the foul smells of human infection.

Also difficult to imagine, was the total ecological system of that
day – the swarms of flies of all species in the manure piles, on the
meat, in the baby's face; maggots in the children's ears, in day-old
meat, in the cattle and hogs; mosquitoes from the moat, worms in

the vegetables and fruit, weevils in the grain, and rats and mice (with their attendant problems) in abundance.

These thoughts were disturbingly ironic since we had gathered there to participate in a symposium on chemophobia. The speakers consisted of a journalist, a psychologist, a physician, three biologists, and two chemists. Although many of us were aware of some people having an excessive fear of chemicals, this was the first that a symposium was held to discuss it publicly. Dioxin was used as the model compound for this discussion. Chemophobia was defined as, "irrational fear of chemicals".

The speakers and their subjects were as follows:

* O. Hutzinger, University of Bayreuth, Bayreuth, Federal Republic of Germany (FRG), "Chemophobia-Introduction of the Session"
* M. J. Boddington, Environment Canada, Ottawa, Canada, "Chemophobia-History of Uncertainty"
* W. Simmler, Bayer AG, Leverkusen, FRG, "Chemophobia-Fear of the Unknown"
* S. Summler, Federal Environmental Agency, Berlin, FRG, "The Problems of Risk Assessment – A Theoretical and Practical Dilemma"
* L. B. Hobson, Veterans Administration, Washington, D. C., "Dioxin and Witchcraft"
* D. E. Klimek, Psychologist, Ann Arbor, MI, "Carcinogens and Chemophobia – Distinguishing Between Ecotoxins and Psychotoxins"
* E. Efron, University of Rochester, NY, "Chemophobia – Pathology in Laymen or Pathology in Scientists?"
* L. Young, Executive Office of the President, Washington, D. C., "Social Controversy and the Dioxin Question".

Although no firm proof was presented, all participants agreed that chemophobia exists. Disagreement came over who is to blame and where the pathology exists.

Perhaps the most credible argument was made by Edith Efron,

the journalist who concluded, "Scientific pathology has bred cultural pathology". The media supposedly had accurately reported information given by scientists. However, scientists had not always represented the data appropriately. One academic scientist, Professor Otto Hutzinger, appeared to agree that the reporting was reasonable but felt the pathology lay with "instant self-appointed experts".

An industrial scientist, D. Walter Simmler, summed it up as follows, "Chemophobia is – justified by industry – formalized by government – nourished by the media – tolerated, at least, by politicians – kindled by whoever expects personal or collective gains of sorts".

A government scientist, A. L. Young, painted the following scenario: "A controversy involving environmental contamination commonly begins with an episode event. Only a few people or animals are exposed, an inadequate sample to determine cause and effect. Reported effects are inconsistent with scientific data and the scientific data are inadequate to confirm or refute all the claims. There is an intense media response followed by an inadequate government response. Special interest groups seize the opportunity to manipulate public and political attitudes. Lawsuits are initiated or threatened. Advisory Groups are set up. These groups almost always reach unsatisfactory resolution. Political action groups demand and obtain congressional action, usually mandated health studies, which often cannot be performed using good science in the time allowed."

Another government scientist, Dr. Martin Boddington, sees all this as equivocation and says, "The equivocation comes from uncertainty. Chemophobia has to do with decisiveness in uncertainty. Ruckleshaus calls it risk management; making decisions in uncertainty but with all facts on the table."

The tragedy of chemophobia is the possibility that an individual may be far sicker from the fear of chemicals than to actual exposure to the chemical themselves. As the psychologist, Dr. David Klimek reminded us, "A promising and fairly new field of research is the area of psychoneuro-immunology, which is the study of the connection between the central nervous system and the psychological process of feelings, perceptions, and inhibitions, and their effects on the immune system."

It is sad to think that scientists ever misrepresent their findings or that self-appointed experts arise who themselves have never done any research. Unfortunately, it is my experience that it is true. Furthermore, at least one judge thought so. In a very interesting decision, Nova Scotia Supreme Court Justice D. Merlin Nann, chastised the scientists testifying for the plaintiffs in a herbicide spraying case in which dioxin was a major part of the debate. He wrote, "There is a noticeable selection of studies which supported their view and a refusal to accept any criticism of them or contrary studies . . . In my view, a true scientific approach does not permit such self-serving selectivity." These were the same scientists who had been feeding stories to the writers of sensational news.

A judge who appreciates good science! That's a rare person.

References

1. O. Hutzinger, editor, "Chemophobia", Chemosphere, Information and News Section, 15, Nos. 9-12, 1986.

2. Pete Earley, "Dioxin Trial Judge Says Witnesses Biased", The Washington Post, October 6, 1983.

3. Fran Munger, "Are Americans 'Healthy Hypochondriacs'?" The Knoxville News-Sentinal, November 23, 1988.

4. Robert C. Cower, "Beware the Excessive Fear of Toxic Chemicals", The Christian Science Monitor, April 16, 1985.

EXPERT DISAGREEMENT

"He was in logic a great critic,
Profoundly skilled in analytic,
He could distinguish and divide
A hair 'twixt south and southwest side",
Samuel Butler, 1612-1680.

1986

In 1986, John Palen, former editor of the Midland Daily News, wrote a paper for a graduate course at Michigan State University in which he correctly attributes much of the difficulty reporters had in discussing the dioxin issue to "expert disagreement".

Newspapers and magazines, according to Palen, handled "expert disagreement" differently. Some reported consensus when there was none, some reported disagreement when there was none, and some reported both sides "side by side" without comment. Publications, attempting to account for "expert disagreement", "did so in several ways: by suggesting that the "other side" was irrational or psychologically disturbed; by suggesting that its science was less competent; or by suggesting that its science was corrupted by politics, economic pressure, or a cynical commitment to corporate power".

Palen's paper also points out that there are certain, "problems in the philosophical foundations of science", "scientific work is an activity deeply influenced by its social nature, and inadequate information introduces uncertainty into many scientific attempts to determine risk".

Palen concludes "that work in the philosophy, history, and sociology of science provides a conceptual framework for the understanding of expert disagreement that is considerably broader than that shown in national press coverage of the Midland dioxin incident". He goes on to say that, "readers have no hope of untangling the issues of expert disagreement if writers do not clearly delineate what these issues are".

There is an element of truth in Palen's writings. There was substantial disagreement among scientists concerned with the behavior of dioxin. The most severe disagreement occurred between those scientists who did research on dioxins (who took their own data, interpreted its meaning, published the work in reputable scientific journals, and attended scientific conferences to openly discuss their methodology and the meaning of their data) and those scientists with prestigious titles who took data published by the first group together with anecdotal reports and re-interpreted it for regulatory agencies and the media.

At the time the dioxin brouhaha began, none of the major scientific disciplines' (analytical chemistry, toxicology, epidemiology, or risk assessment) need for the assessment of its impact at parts per billion levels had a philosophical basis well enough developed to accomplish the task. To measure quantities less than a part per billion, analytical science had not reached a consensus on matters of detection, determination, certainty, interferences, false positives, false negatives, etc. Data was being generated by laboratories with different attention to these matters and the resulting data sets were not comparable.

Toxicology had proven that animal testing was an excellent way to predict acute effects of high concentrations of chemicals on humans. But when faced with assessing the toxicity of parts per trillion amounts to humans thirty to forty years after exposure, concepts such as "no effect level" were placed in doubt.

Epidemologists had been highly successful in identifying and characterizing diseases and acute toxic exposures where the number of patients affected were large. But the challenges of the effects of toxins on a few at a very low concentration presented new questions not previously considered.

Risk assessment depends on high quality data and interpretation from analytical science, toxicology, and epidemology. Since, in the beginning, none of the three provided unquestionable information, "experts" made their own interpretations, often resulting in amazing conclusions, almost always scary, and often wrong.

Perhaps the best support for my conclusions based on personal observation and experience comes from a survey of 1300 health professionals in the fields of toxicology, epidemology, medicine, and other health sciences. Commissioned by the Institute for Regulatory Policy, the 1991 survey shows that 81 percent of those surveyed believe that public health dollars for reduction of environmental health risks are improperly targeted; 87 percent agree that it is impossible to calculate human cancer deaths accurately based solely on extrapolations from animal data, and the data overwhelmingly support a "weight-of-the-evidence" approach that takes all plausible human and animal data into account.

As a result of the study, the Institute wrote, "Toward Common Measures. Recommendations for a Presidential Executive Order on Environmental Risk Assessment and Risk Management Policy". This included guidelines and principles for risk assessment and management which would require consistent performance and regulations among various agencies. Although I generally abhor presidential executive orders, I supported this one because any action which may persuade scientists to reach consensus is welcome.

In private discussion, Dr. Palen appears firm in his belief that the press had covered the dioxin issue very well. However, in an article in the March 15, 1992, issue of The Midland Daily News, he is quoted as saying, "The threat of dioxin was to social cohesiveness, responsible capitalism and moral leadership, all values that underlie media coverage in the United States. Ecologism challenges the media's believe that human beings should be the central concern. The press, in covering science and technology, needs to expand its focus from humans to the environment as a whole".

No one has, to my knowledge, been able to develop a workable formula or system for resolving "expert disagreement". A gallant effort was made by Milton R. Wessel who championed "rule of reason" and "dispute resolution" workshops and conferences to achieve consensus among "experts". Those on 2,4,5-T, in which I was involved, appeared to be successful. However, the next time the news reporters quoted "experts" who had been in attendance, the same old scary inaccurate statements were made. Thus the self-appointed "experts" remained tied to their original opinions.

References

1. Milton R. Wessel, "Rule of Reason", Addison-Wesley, 1976.

2. Milton R. Wessel, "Science and Conscience", Columbia University Press, 1980.

RETURN TO THE MOUNTAIN

"Our civilization is still in the middle stage, no longer wholly guided by instinct, not yet wholly guided by reason", Theodore Drieser, 1871-1945.

1988

It is 1988 and I have just returned from another visit to the Crummett Mountain community. The mountains, hills, and ridges are still there. So are the rocks. The same springs are still flowing and the creeks babble in roughly the same beds as before, but now they flow through culverts and under bridges. The banks that were clay are still clay. Those that were limestone are still limestone. Sand deposits are still sand deposits.

The geological features still bear the names given them by ancestral pioneers in the eighteenth century. The creeks are still called Crummett Run, Black Thorn Creek, The South Fork, Little Fork, Brushy Run, Stoney Run, etc. Some of these names are really interesting. For example, there is the Little Fork of the South Fork of the South Branch of the Potomac River. Simmons Mountain, Hoover Mountain, Bother Knob, Reddish Knob, and Shaw's Ridge are still there. However, the Knobs have lost their fire towers. Today, forest fires are spotted by helicopter or satellite.

Landmark buildings still bear the same names. The post offices are the same; Sugar Grove, Headwaters, Doe Hill. The churches still have the same names—Crummett Run, Wilfong, Thorn Chapel, Brushy Fork, Moatstown, Trinity, St. Paul, and St. John's. Sadly, the one-room schoolhouses are gone. Only one, Sinnett's Lane, remains.

The split-rail fences are gone, consumed by decay. Wire fences have replaced them. Wire too has replaced the board fences around

Crummett Run Church of the Brethern.

meadows and picket fences around lawns. Many barns have disappeared. Swinging foot bridges can still be seen in some areas.

Farm practices have changed. No longer can one find fields of corn, wheat, rye, buckwheat, or oats. One only sees sheep and cattle grazing in the fields and huge poultry houses. Horses are seldom seen. No longer does one see flocks of chickens, turkeys, ducks, geese, or guinea fowl roaming free. The turkey that gobbles in the wood is wild. Deer herds abound, moving over the old homestead like a host of shadows, consuming small flora so that many of the old familiar small plants are no longer easily found.

Barefoot boys with hoes can no longer be found. They are extinct. Stubbed toes, callused feet, bruised shins, and blistered hands are not permitted. Straw hats and sunbonnets are not in evidence. Only a very few children reside here. Those who do, wear shoes except when at the beach. The beaches are artificial, produced by the damming of mountain streams. This was done, the Federal government says, to provide flood control. However, when floods occur, the major concern of the people is the possible bursting of these dams.

The old gristmills, owned by individuals, are seldom seen. The wheels are rusted. The dams have filled with silt. The buildings are falling down. Sawmills, owned and operated by individuals or small groups, can no longer be seen. Wood-fired steam engines no longer exist. Many of the wood burning, potbellied stoves have been replaced by oil fired furnaces. Kerosene lamps have been replaced by electric lights. Most houses have hot and cold running water. Restrooms have been installed in the church.

Boys, barefoot or not, are no longer needed to destroy the weeds and kill the bugs. The raising of crops is now done on large farms located far away. The development of chemical pesticides allowed these things to be done much more economically on large farms using large mechanical equipment to cultivate and harvest. Thus, the people have more leisure time.

At my mother's funeral, I shook hands and hugged hundreds of relatives and friends. Only one or two of them had callused hands, which was one of these peoples' most prevalent conditions in the past. Much of the past hardships had vanished.

Although still living on the land, the people are not as dependent on natural events as before. Weather conditions are less worrisome. Insect and weed infestations are no longer viewed as a plague. The folk don't blame God so much for their problems. They merely blame the government. They no longer have much to say about ghosts, spooks, and witches. People still remember persons who were witches, but they know of none still living.

DECADE VIV
ON REFLECTION

THE WIZARDS

"We live at the edge of mystery . . .", J. Robert Oppenheimer.

"Those damned Dow guys present their work with complete confidence. They seem to be daring us to find anything at all wrong with it and you know—we can't!" The eminent environmental scientist drew his mouth into a straight line and listened as the group around him agreed. (They then referred to us as "The Dow School".) This scene occurred on October 22, 1980, during a coffee break at the Institute Superiore di Sanita, Rome, Italy, in the workshop on "Impact of Chlorinated Dioxins and Related Compounds on the Environment". The speaker referred to Lester L. Lamparski and Terry J. Nestrick, each of whom had just finished presenting a paper on the analytical chemistry of chlorinated dioxins and the separation of the 22 isomers of TCDD. Similar remarks were to be made later in many scientific conferences and university lecture halls where these two scientists, often called, "The Wizards" spoke.

Adding to the discomfort of most potential critics, the Wizards always appear in t-shirt and blue jeans. The shirts bear designs and sayings, which they deem appropriate for the occasion. Visitors are often shocked and intimidated by the messages. Adding to the image are the black mustache and Apache hairstyle, complete with headband, of Nestrick and the red bearded long hair Viking style of Lamparski. They are sometimes judged to be anti-establishment and once were ostracized from an exclusive restaurant in Canada, while there at the invitation of the Canadian government.

Some Dow managers were embarrassed by the title, "The Wizards", and requested that supervision ban the use of the term. Try as they might to eradicate it, however, the title was continued to be used by university professors and prestigious institutions. Nestrick and Lamparski have continued to receive mail such as that addressed to "The Wizards, Dow Chemical, Analytical, 574 Building, Midland, MI, 48640", from the National Bureau of Standards. They have also received telexes from well-wishers such as the one addressed simply as "The Wizards, Analytical Labs, Midland", from the TOD, dated May 21, 1984.

Julie Morrison, a reporter for the Midland Daily News, described them as "outspoken, freely criticizing everything from the government to their own employers. And they are independent, having on several occasions stood by theories the remainder of the scientific community snubbed". She also pictured them as arrogant. "They agree they are the best research team in the country."

Donald G. Barnes, staff director of EPA's Science Advisory Board, wrote, in a trip report after visiting Dow on September 8-10, 1981, as follows: "Met with Frank Knoll and The Wizards, Lamparski and Nestrick.

The latter are delightful products of a slightly irreverent anti-establishment world, which was probably broken sometime in the early 1970's. In speaking with them I get the impression that one believes that if it can't be stated on a t-shirt, it's not worth saying; with the other, it's a matter of hyperbole and a tongue that is often in cheek. Basically, they are an entertaining and remarkably creative and capable pair. They have accomplished amazing feats of chemical analysis and organic synthesis in the area of dioxins."

Although these characterizations are rather accurate, they are not the whole story. Further insight can be gained by the response given by Nestrick to the question, "Why do you strive so hard for perfection?" posed by William "Bill" Lowrance of the Rockefeller University. Terry said, "Because it is like training for the Olympic games. We have to be the very best to win."

After release of the "Trace Chemistries of Fire" data, the Wizards were invited to go on a lecture tour of some major universities. The following description of their visit was received from one of the professors.

"On behalf of all of the faculty and students in the Analytical Division, I want to express my deepest appreciation for

Dr. Lester Lamparski keeps his mass spectrometer clean and
ready.

Dr. Terry Nestrick always has his chromatography column
packed just so.

the generosity of Dow Chemical and particularly yourself in

making it possible for Les Lamparski and Terry Nestrick to visit here. Although the excellence of their analytical chemistry should not have been a total surprise to me, their visit here significantly strengthen my already high admiration of these two unusual professionals. Their presentation of the work on the TCDDs, as well as the other aspects of their visit among us, had an electrifying effect on the entire group. The grasp of what could be accomplished within the framework of a forward looking industrial organization made what I believe to be a highly positive and long-lasting impression on every one. The Wizards failed to hide their generally unqualified enthusiasm for their employers and associates. It was beautiful to see that pride of craftsmanship is alive and well at Dow Chemical and, by reasonable extension, in American industrial analytical laboratories. It is also obvious and deeply indicative of Dow's leadership that you put substance before form and productivity before conformity."

At another university, an advocate group had planned a protest against Dow. The leader attended the lecture and was prepared to signal the start of a demonstration. However, he left early and was heard to say, "Those scientists are for real. No need for my heckling".

More recently, the reviewer of a manuscript entitled, "Compound-specific HRGC-LRMs Determination of Halogenated Dibenzo-p-dixons and Dibenzofurans in Environmental and Biological Matrices" to be published by the World Health Organization (WHO) commented: "The work of Lamparski et al., is first rate. The Dow scientists pioneered much of the analytical work relating to determination of chlorinated dibenzo-p-dioxins (CDDs) and dibenzofurans (CDFs) in the environment and they are still leaders in this area. Their methods have been adapted or modified by many other laboratories and in several round-robin investigations, the Dow methods have proven to be among the most effective."

We unleashed the Wizards to work on the dioxin question in 1976. Since then, the following have been accomplished.

1. They were the first to discover the presence of 2,3,7,8-

TCDD at the part per trillion level in fish from the
Tittabawassee River.

2. They were the first to separate all 22 isomers of tetrachloro-
dibenzo-p-dioxins and all 109 isomers of
hexachlorodibenzo-p-dioxins.

3. They were the first to determine 2,3,7,8-TCDD isomer
specifically.

4. They were the first to find the chlorinated dioxins in soil
and dust from cities other than Midland.

5. They were the first to find chlorinated dioxins in automobile
mufflers, cigarette smoke, and wood soot.

6. They were the first to find chlorinated dioxins in National
Bureau of Standards urban dust standard sample

7. They were the first to apply quality control criteria together
with a thorough understanding of each step in the analytical
process to allow meaningful data to be obtained from a non-
statistically designed study.

8. They were the first to find chlorinated dioxins in a sealed
sample of sewage of municipal sludge which had been
exhibited at the 1933 Chicago World's Fair.

9. They were among the first to prove "de novo" synthesis of
the chlorinated dioxins in fire.

10. They supplied the data required for the development of
"The Trace Chemistries of Fire" hypothesis and that for
developing mechanisms which explain how it is reasonable.

11. They were the first to analyze water for TCDD at a part per
quadrillion level with repetitive reliability sufficient for the
regulation of effluent from a industrial site.

In September 1988, the Wizards were awarded doctor degrees
by the University of Umea, Sweden. Under the direction of Professor
Christoffer Rappe, they have earned these by:

1. Each passing comprehensive examinations in the fields of
mass spectrometry and nuclear magnetic resonance.

2. Together doing a collaborative study of Baltic Sea sediments

with Professor Rappe and his other students.

3. Each writing a thesis incorporating all the work they had done on the dioxins.

4. Together teaching liquid chromatography to Umea University students.

5. Individually defending their theses before adversaries from countries other than Sweden or the United States.

The announcement elicited a proclamation from Professor Henry Freiser of the University of Arizona as follows.

PROCLAMATION
PERIOD OF MOURNING FROM 11/88 TO ??

INCLUDING: 1)GNASHING OF TEETH
2)WAILING AND HOWLING
3)RENDING OF CLOTHING
4)COVERINGS IN SACKCLOTH &
ASHES (2378-FREE?)

OCCASIONED BY: DEMOTION OF
LES LAMPARSKI
AND
TERRY NESTRICK

FROM THE EXALTED STATE OF
WIZARD OF CHEMISTRY
TO LOWLY DENIZENS OF ALREADY CROWDED
RANKS OF Ph(ooey)D

LAMENT!   FRIENDS, RELATIVES & INNOCENT
BYSTANDERS –
WHERE DID WE GO WRONG?
WHAT'S TO BECOME OF THEM NOW?

On January 5, 1989, Drs. Nestrick and Lamparski were presented with the Sigma Xi Award for the best published paper in the local Sigma Xi Chapter. This was the second time they had won this award.

References

1. Julie Morrison, "Wizards at Work", Midland Daily News, March 12, 1984.

2. L. L. Lamparski, T. J. Nestrick, R. H. Stehl, "Determination of Part-per-Trillion Concentrations of 2,3,7,8-tetrachlorodizenzo-p-dioxin in Fish", *Anal. Chem.*, *51*, 1453, 1979.

3. L. L. Lamparski, T. J. Nestrick, "A Comprehensive Procedure for the Isomer-Specific Determination of Tetra-, Hexa-, Hepta-, and Octachlorodibenzo-p-dioxins at Parts-per-trillion Concentrations: Particulate Samples", *Anal. Chem.*, *52*, 2045, 1980.

4. T. J. Nestrick, L. L. Lamparski, D. I. Townsend, "An Approach to the Identification of Tetrachlorodibenzo-p-dioxin Isomers at the 1 Nanogram Level Using Photolytic Degradation and Pattern Recognition Techniques", *Anal. Chem.*, *52*, 1865, 1980.

5. L. L. Lamparaski, T. J. Nestrick, "Synthesis and Identification of the 10 Hexachlorodibenzo-p-dioxin Isomers by High Performance Liquid and Packed Column Gas Chromatography", *Chemosphere*, *10*, 3, 1981.

6. T. J. Nestrick, L. L. Lamparski, "Isomer-Specific Determination of Chlorinated Dioxins for Assessment of Formation and Potential Environmental Emission from Wood Combustion", *Anal. Chem.*, *54*, 2292, 1982.

7. L. L. Lamparski, T. J. Nestrick, V. A. Stenger, "Presence of Chlorodibenzodioxins in a Sealed 1933 Sample of Dried Municipal Sewage Sludge", *Chemosphere*, *13*, 361, 1984.

8. W. B. Crummett, T. J. Nestrick, L. L. Lamparski, "Analytical Methodology for the Determination of PCDDs in Environmental Samples: An Overview and Critique". Chapter 6: "Dioxins in the Environment". Michael A. Kamrin and Paul W. Rodgers, editors. Hemisphere Publishing Corporation, New York, 1985.

THE ULTIMATE AWARD

"Think where man's glory most begins and ends,
And say my glory was I had such friends",
William Butler Yeats, 1865-1939.

18 March 1988

When the Wizards were introduced at my retirement dinner, the Great Hall roared with laughter. It seemed as though the Knights of the Round Table had reassembled in King Arthur's Court. Above the din the voice of Henry Freiser was heard, "Oh! My God! They have ties on!"

Dr. Nestrick quickly took charge. After pleading that any stammering he did was to avoid using his usual four-letter words, he continued, "Tonight you have seen Warren honored by a number of awards, but now we are about to give the ULTIMATE AWARD. Tonight we are somewhat surrounded by people—renowned scientists such as Rogers and Freiser. Between them, Buck and Henry, they have won every important award save one—one they will probably never win.

First, it is the oldest award in the field of chemistry. It dates all the way back to alchemy. Second, it is the only award with historical documentation in the lay literature. Third, it is extremely exclusive. It is awarded to but three individuals per century. This is the last one in the 20th century.

I present to Warren Crummett this impervious glass figurine, 'The Wizard', giving you New Rank in recognition of your services and the last 12 years goading Les and me to death!"

I was speechless! The Great Hall roared! For an instant we were back in time.

The figurine was hand blown of different colored glass. The white bearded face of "The Wizard" with bright eyes was topped by a black pointed hat, speckled with stars. Dressed in a flowing black robe with large transparent sleeves, "The Wizard" extends his left hand holding a fire complete with red flame and white smoke. The right hand is raised and extended in a gesture of peace.

It was a very satisfying experience for, if I had really goaded Les and Terry "to death" over the last 12 years, we had been at equilibrium all along.

The Wizard Award!

It was time for some reflection, I thought of our last publication together, "Thermolytic Surface-Reaction of Benzene and Iron(III) Chloride to Form Chlorinated Dibenzo-p-dioxin and Dibenzofurans", which won the Sigma Xi Award for "the most outstanding published research" in the Midland area over the past three years. It, together with papers from leading scientists in Europe, provided proof of the *de nova* synthesis of chlorinated dioxins. In other words, it had now been shown that chlorinated dioxins could be formed from various inorganic chlorides and hydrocarbons in a fire. My name was on that paper only because I had to debate, at great length, the scientific and political implications of its publication with Dow management. Management had determined that it was not profitable for a company to be "on point" on any environmental issue. Now publishing by any industrial scientist was once again impeded by their own management, not by editors of journals. In the 1970's many outside Dow seemed to think our research had all the answers on dioxin and we were not sharing the information. In truth, we had shared everything. Unwarranted attacks on the company, however, now placed the privilege of publishing in jeopardy. Extreme activism often produces the opposite of the intended.

References

1.H. Hagenmaier, M. Kraft, H. Burnner, and R. Haag, "Catalyic Effects of Fly Ash from Waste incinerator Facilities on the Formation of Polychlorinated Dibenzo-p-dioxins and Polychlorinated Dibenzofurans", Environ. Sci. Technol., 21, 1080, 1987.

2.T. J. Nestrick, L. L. Lamparski, W. B. Crummett, "Thermolytic Surface-Reaction of Benzene and Iron(III) Chloride to Form Chlorinated Dibenzo-p-dioxin and Dibenzofurans", Chemosphere, 16, 777, 1987.

3.L. Stieglitz, G. Zwick, J. Beck, W. Roth, H. Vogg, "On the De-Nova Synthesis of PCDD/PCDF on Fly Ash of Municipal Incinerators", Chemosphere, 18, 1219, 1989.

SOME GREAT ANALYTICAL PIONEERS

"And now the matchless deed's achieved, determined, dared, and done", Christopher Smart: Song to David.

If the name "The Wizards" was useful for Nestrick and Lamparski, names from Medieval times would also fit other analytical scientists, including myself, involved in searching for dioxin. Whether from academia, government, or industry, analytical scientists are alike in their vigor to collect sound data, determine the facts, and develop sound hypotheses. Almost universally, these scientists place their loyalty first in science itself, secondly to their peers, and last to the institution that employs them. However, they approach the gathering of information and the teaching of others in strikingly different ways. These different approaches add tremendously to the strength of scientists to develop the truth, although often they appear to be wrangling unnecessarily.

Careful observation of these different approaches reveals that they are very similar to those of different characters of Medieval times. It is tempting for me to bestow selected fantastic titles of king, queen, baron, count, earl, lord, knight, bishop, cardinal, pope, pilgrim, page, emperor, monk, nun, crusader, peasant, archer, friar, jester, lady, herald, princess, prince, conqueror, etc., on the various analytical scientists involved it the dioxin saga.

If that were done, certain ones would merit the highest rank. One of these is David Firestone of the U. S. Food and Drug Administration. Dr. Firestone developed the extraction techniques and many of the chromatography systems used for the clean up of fat for the determination of dioxins. In the 1960's, quantitative mass spectrometry was unavailable to his laboratory. So he adapted microcoulometric chloride and electron-capture detection methods for gas chromatography to detect and measure the chlorinated dioxins. In 1970, I visited his laboratory and learned much from our discussions of solvent systems and partition coefficients.

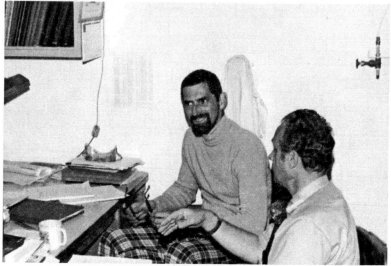

Dr. David Firestone of FDA visits a delighted Dr. Lew Shadoff.

While there, I noted that he had analytical standards of hexachlorinated dibenzo-p-dioxins. I asked how I could obtain samples. He got some clean vials and poured some of his standards into them, labeled them, and handed them to me. "What should I do with these?" I asked. "Why not stick them in your coat pocket and take them home?" he replied. Some of these standards were later used by the Wizards to prove the separation of the ten isomers of hexachlorodibenzo-p-dioxin.

Tom Tiernan is another early dioxin analytical chemist of highest merit. Dr. Tiernan was still with the U. S. Air Force when I first met him, but quickly was made a professor at Wright State University. He was one of the relatively few who was experienced enough in quantitative mass spectrometry to conduct experiments and interact appropriately with Lew Shadoff. Early, he recognized the inherent difficulties in trace analysis and took appropriate steps to control them. Furthermore, he established one of the first contract laboratories capable of analyzing samples for both government and industry. On March 24, 1988, Tom drove 600 miles to participate in my retirement dinner.

Bob Harless of the U. S. Environmental Protection Agency is another mass spectroscopist who not only produced excellent data, but kept the agency honest in its assessment of the source and quantity of dioxin. He developed a set of criteria for determining whether a signal was due to TCDD or some interference. This was a major contribution because, for the first time, analytical chemists from various laboratories could agree on data. Bob is a man of the highest principles and integrity.

Bob Harless, EPA's world class mass spectroscopist with instrumentation.

Aubrey Dupuy of the U. S. Environmental Protection Agency did the separations for all the mass spectroscopists in the early work and for Bob Harless for many years. He brought a credibility to the agency which was very important. Calm and always unruffled, he often soothed even the most belligerent of EPA attorneys. He taught the meaning of the chain of custody of samples and the precautions necessary to avoid contaminating the samples.

Michael Gross of the University of Nebraska was encouraged by Lew Shadoff to become a member of the Dioxin Implementation Plan. His expertise in high resolution mass spectrometry was needed if the experiment was to be successful. His group obtained some of the best data the Plan generated, making the information more credible. Since he analyzed samples for those who could not afford to pay, he was often involved in public issues. Later he was brought to Dow by Lockhart "Buck" Rogers as a member of the Blue Ribbon Panel to assess the merits of the analytical chemistry used to develop the "trace chemistries of fire" hypothesis. Mike did not believe in the hypothesis, however, and called to warn me that the academic scientific community would not accept it. I informed him that the evidence was mounting that it was correct. The latest data showed that dioxin was present in a 1933 sample of Milorganite, a commercial fertilizer made from sewage sludge in Milwaukee. He was very upset because he had just put this fertilizer around his roses, and was afraid he had poisoned himself. He quickly recovered, however, when he recognized that the scary stories in newspapers had temporarily clouded his scientific judgement.

I first met Christoffer Rappe on June 24, 1978, at an international conference entitled, "Health Effects of Halogenated Hydrocarbons" held by the New York Academy of Sciences. He spoke on, "Dioxin, Dibenzofurans and other Polyhalogenated Aromatics: Production, Use, Formation, and Destruction". The presentation was powerful, and he spoke convincingly as one having authority. I was greatly impressed. I was saddened, however, to note that although Dow scientists had contributed extensively to the dioxin literature, only one Dow reference was included. Later, my report on the "trace chemistries of fire"hypothesis failed to include Rappe's work on pyrolysis, a related phenomena. These apparent indifferences to each other's work caused us to appear to be diametrically opposed to each other's position. However, this was not true and our views on dioxin are now very close. A nice working relationship has been achieved. I have the highest regard for Professor Rappe's accomplishments, both as an organic synthesis chemist and as an analytical chemist. His conclusions are almost always based on measurements his own group has made.

I first encountered Professor Dr. Otto Hutzinger at "The Ninth Annual Symposium on the Analytical Chemistry of Pollutants" at Jekyll Island, Georgia, May 8, 1979. He, together with K. Olie and P. L. Vermeulen, whom I met later, had in 1977, held it to be possible that chlorinated dioxins and dibenzofurans be formed by *de novo* synthesis. Frankly, I did not believe this and when other Dow scientists invited him to visit Dow, I kept in the background. The Dow data generated in the investigation leading to the "trace chemistries by fire" hypothesis forced me to rethink my position. Thereafter, I have used every opportunity to work closely with him. He understands the big picture of environmental relationships. He has proven himself to be a great organizer of symposia and an extraordinary keynote speaker on dioxins, always bringing a fresh viewpoint. Simple, but highly meaningful, experiments are his trademark.

Elsewhere in this book I have discussed the great work of Bob Baughman and Matthew Meselson. When Dr. Baughman graduated from Harvard and sought refuge in medical research, Patrick O'Keefe was hired to finish the contract work on dioxin. Pat, a native of Ireland, had studied in Oregon. His demeanor as a member of the Dioxin Implementation Plan, kept us alert and caused me to jokingly wonder about his possible relationship to the Blarney Stone. Now, as a part of a fine laboratory of the New York State Department of Health, he is doing a great job.

Host Prof. Y. Masuda welcomes Drs. Chris Rappe, Helle Tosine,
Ray Clement and others to the dioxin conference in Fukouka,
Japan.

Helle Tosine set up an excellent laboratory at Ontario Ministry of
the Environment for the determination of dioxins and dibenzofurans. In
this laboratory, she included the best facilities proven useful by laboratories
        earlier in this type of work. I consider her effort a major achieve-
ment. Within this laboratory is an outstanding analytical chemist, Dr.
Ray Clement. He has also represented the province of Ontario extremely
well. He is probably the most free-speaking analytical scientist in government anywhere, but more importantly, he writes extremely well. I
often seek his advice.
        The team of Richard Hummel and Lewis Shadoff of The Dow
Chemical Company were the first to make the determination of
2,3,7,8-TCDD truly quantitative on a repetitive basis with all the
quality assurance parameters built in. They, together with Rudolph

Stehl, were the first to use gas chromatography to introduce the sample into a high-resolution mass spectrograph and prove that quantitative information can be obtained. Elsewhere in this book, I have described some of Dr. Shadoff's incredible performances.

Believe me, if I were a head of state, or president of a prestigious scientific society, all of these scientific giants would be given many major awards. Their achievements have set world records in analytical science. The fact that few of these have been appropriately recognized by their scientist peers casts shadows on the award given to others and brings to questions the objectivity of other scientists.

Working with these exceptional people was a challenge but a great experience for me! I miss them greatly! Occasionally I see Matt Meselson on a television talk show as a scientist consultant. He is a true advocate for science.

THE BEAT GOES ON

"Believe those who are seeking the truth; doubt those who find it."

1988-2000

Emboldened by their success in getting governments to regulate dioxin at absurdly low levels, environmental extremists have called for the banning of chlorine and all chlorine-containing compounds. This impossibility is necessary, they say, because "the cumulative damage these compounds do to human health may be incalculable". This quoted statement is irrefutable. If the statement were made about oxygen-, nitrogen-, or sulfur-containing products, it would still be irrefutable. Therefore, it suggests that those who portray environmentalism as a new religion are correct. Nasty, scary things can be said about any molecule or group of molecules.

According to Robert H. Nelson, this type of environmentalism is based on: "a sense that modern civilization tempts man to do evil and represents retrogression rather than progress: an apocalyptic foreboding concerning ecological catastrophe and the near-term future of the earth; an attitude that human reason, as today embodied in science, offers false promises and alienates man from his true self; a view that a widespread sinfulness has infected the world . . . and a desire to return to an earlier and more natural existence – the existence of the Earth of long ago, in which the products of modern science and economics would be banished".

It is certainly the right of individuals to embrace environmentalism as a religion. However, this fanaticism is dangerous since environmental laws now often arise more from this religion than from science. Thus congress could reasonably be accused of passing laws establishing a religion. These possibilities have been brilliantly discussed by Cathy Young in, "Worshipping at the altar of environmentalism". Unfortunately, my experience agrees with her assessment. From my perspective it appears that much of the news media has joined this worship. It is a very scary thought.

I have an extreme aversion to the chlorine molecule. Ever since I first smelled it in a college laboratory of general chemistry, I have tried to avoid it, even to the point of refusing to synthesize it for college credit. Later, working in a chlorine manufacturing plant in Virginia, I breathed a little daily and once got enough to require medical treatment. But, I am totally persuaded that, without chlorine to treat municipal drinking water, I would not have survived the transition from drinking water from the free-flowing Shenandoah Mountain springs to drinking municipal water in college and industrial towns.

In the early 1900's typhoid fever, diphtheria, malaria, cholera and related diseases were common in the eastern United States and were spread by contaminated water supplies as well as mosquitoes. Epidemics were so severe that some thought that the Horace Mann slogan of, "Go west young man", was as much for health reasons as for wealth. These diseases rarely attacked people living high on the hills and mountains and drinking fresh spring water. But, not being exposed they didn't develop immunity to these maladies. Thus my mother lost her father to typhoid fever and two brothers, while they were still toddlers, to diphtheria.

By 1934, when I started high school in a town with a common water supply, chlorination made the water safe for me even though I had not been immunized for any disease other than diphtheria. I was safe through college, graduate school, and an industrial career although I could never be made immune to typhoid fever. So I was and am well aware of the enormous contribution chlorine has made to society.

My career with chlorine has made it clear that this molecule can and is being safely managed. It can be allowed to react in ways to benefit society and prevented from reacting in ways to harm society. Even when human error allowed a dramatic thing to occur, the chlorine molecule behaved in totally predictable ways. For example, in

1944, when I was away on vacation from the Sovay Process Company, a tank car being loaded and containing about 10 tons of liquid chlorine, was accidentally struck by a railyard locomotive and disconnected from the loading facility. Its content boiled out and, being heavier than air, rolled as a dense cloud of vapor along the ground down to the James River where it was hydrolyzed. This behavior had been predicted and communicated to workers. Those workers who had believed the prediction and acted accordingly were only slightly exposed. Those who panicked were briefly hospitalized.

I do not enjoy breathing molecules of chlorinated hydrocarbons. However, the methylene chloride, chloroform, carbon tetrachloride, methyl chloroform, trichloroethylene, perchlorethylene, chlorobenzene, and other similar chlorinated hydrocarbons were very useful as extraction solvents in analytical chemistry. I used these solvents in various glassware, especially separatory funnels, in the open laboratory. I thus breathed vapors of these materials on almost all my work days over a period of at least 25 years.

Those who fear these molecules must marvel greatly that I am still alive at 80 years of age. In addition to exposures already described, I must also confess that cough syrup with chloroform was the best remedy for congestion when I was a child and I continually suffered from coughs and colds. As a scientist, I report this experience with reluctance and know that it proves nothing. However, anecdotal information designed to create fear is given top billing by the news media. So, common sense observations attesting to the innocuous effects of certain molecules cannot be ignored.

One of the most recent scares concerning chlorinated hydrocarbons is based on allegations that these molecules are environmental estrogens. This activity has been "linked" to "abnormally small penises" in Florida alligators, an increase in the chances of a woman developing breast cancer, a decrease in the sperm count in men, an abnormal thyroid function in almost everything that swims or flies, and an increase in female offspring. Later, more comprehensive and scientific studies have refuted the women's breast cancer and men's sperm count scare. More complete and definitive studies will likely refute the other allegations as well. If so, the alarmists will not stop. Other biological "links" will be made as the beat goes on.

Typically, the Environmental Protection Agency has leaked the preliminary results of a new risk assessment of dioxin in which the claim is made that "subtle effects" on fetal development and the immune system are more important than the risk of developing cancer. The conclusion is based on new mathematical assumptions" that have not been peer reviewed. More regulation will be attempted.

Such regulation will cost billions but will have no effect on human health. Proper management of dioxin sources has already reduced the "dioxin" level in the average American from 18 to 5 parts per trillion according to Dr. Larry Needham, a credible scientist with the Center for Disease Control and Prevention in Atlanta. I estimate that the levels will continue to decline under the present management to about 1 part per trillion for the average American and will remain at about that level. Further bans or restrictions on industry will not affect this level significantly.

Although the Society of Toxicology, the American Medical Association, and the American Chemical Society have all characterized the chlorine ban as unscientific, advocacy groups such as Greenpeace still rant and rave in favor of such a ban. The beat goes on. The beat will go on so long as EPA justifies expanding budgets by "tweaking science" and "hijacking the anxiety of a technically naive public with insinuations of risk, and then regulating on the pretext that people are afraid". This according to Gis Batta Gari, Director of the Health Policy Center in Bethesda, MD.

References

1. Robert H. Nelson, "Environmentalists and the New Gospel", Chicago Tribune, August 14, 1990.
2. Cathy Young, "Worshipping at the altar of environmentalism", The Detroit News, January 6, 1991.
3. Doug Henze, "Merits of chlorine topic at conference", Midland Daily News, November 30, 1992.
4. Bette Hileman, "Concerns Broaden over Chlorine and Chlorinated Hydrocar-

bons", C & E News, April 19, 1993.

5.Bradley R. Lienhart, "More on Chlorine", Letter to the Editor, C & E News, May 10, 1993.

6.Michael Gough, "More on Chlorine", Letter to the Editor, C & E News, May 10, 1993.

7.Bette Hileman, "Debate Over Phaseout of Chlorine, Chlorinated Organics Continues", C & E News, December 6, 1993.

8.Steve Waring, "Chlorine use up for debate", Midland Daily News, January 19, 1994.

9.Steve Waring, "EPA plan aims at chlorine", Midland Daily News, February 3, 1994.

10.Bette Hileman and David Hanson, "Curbs on Chlorine Sought", C & E News, February 7, 1994.

11.James W. Crissman, "Chlorine a necessary part of all of us", Midland Daily News, February 10, 1994.

12.Michael Heylin, "Time Out", C & E News, February 14, 1994.

13.M. C. Carpenter and Jorge L. Cerame, "Chlorine ban ill-advised", Midland Daily News, February 16, 1994.

14.Charlie Cray, "Greenpeace speaks out against chlorine", Midland Daily News, February 27, 1994.

15.Enrique Sosa and Larry Washington, "Speak out on chlorine issue, before its too late", Midland Daily News, February 27, 1994.

16.James W. Crissman, "Myths, science, politics: Altruism not in anybodies job description", Midland Daily News, February 27, 1994.

17.Bette Hileman, "New approaches to toxic chemical control urged", C & E News, February 21, 1994.

18.John Catenacci, "Chlorine ban issue merely a shot across the bow", Midland Daily News, March 13, 1994.

19.Eric Larsen, "Banning chlorine the 'second grand experiment'", Midland Daily News, March 13, 1994.

20.Vernon A. Stenger, "Essential Chlorine", Detroit Free Press, March 15, 1994.

21.Raymond W. Luckenbaugh, "Sperm count speculation", C & E News, March 21, 1994.

22.Bill Schomer, "Flawed scientific reasoning", C & E News, March 28, 1994.

23.Gordon W. Gribble, "Organochlorine drugs", C & E News, April 18, 1994.

24.James W. Crissman, "Two of Ecoalarmists' fears can be dismissed", Midland

Daily News, May 1, 1994.

25. Keith Schneider, "Dioxin conclusion sparks storm of dissent", Midland Daily News, May 11, 1994.

26. Bette Hileman, "Toxicologists Criticize Proposed Chlorine Ban", C & E News, October 17, 1994.

27. Cheryl Wade, "Greenpeace launches 'case against chlorine'", Midland Daily News, September 16, 1995.

28. Doug Heenze, "Greenpeace Plan now for chlorine phaseout", Midland Daily News, September 17, 1995.

29. C & E News, "High exposure to dioxin linked to increase in female offspring", August 19, 1996, p 34.

30. C & E News, "Dioxin to be listed as known carcinogen", November 17, 1997, p 26.

31. Joseph H. Hebert, "Draft: Dioxins cancer risks may be greater than previously thought", Midland Daily News, May 17, 2000.

32. Editorial, "Dioxin News: A question of scale", Midland Daily News, May 24, 2000.

33. Beth M. Bellor, "Dioxin report due today", Midland Daily News, June 12, 2000.

34. Gis Batta Gari, "Turning point for EPA?", Washington Times, June 23, 2000.

BRIDGEWATER FAIR

"Bridgewater fair,
My heart's sweet care",
John Wayland.

1991

To my amazement I was invited back to my college alma mater as a visiting lecturer for the spring trimester of 1991. The intent was to inform the students about analytical science and its contribution to modern society. Lectures and discussions were planned on the role of analytical science in everyday life, environmental studies, industry, government, and academia; academic requirements and continual education for a career in analytical science; the various roles in analytical science; the philosophy of analytical science and industry; the role of government in industry; good laboratory practices and quality assurance, etc.

I found it difficult to talk about any of these subjects without citing dioxin as an example. On doing so in my second lecture, the attendees urged me to talk about dioxin and work the other subjects into the discussions. So the rest of the sessions were conducted in that mode. The participants, consisting mostly of chemistry professors, retired professors, and chemistry majors now entered the discussion in a much more vigorous way, bringing articles and books to the lectures to get my reaction and test my integrity. It was fun and a great interaction.

I was able to weave into the sessions a sense of the philosophical, psychological, and personal college experiences, which were critical to my future involvement with dioxin. For example, coming from a family financially devastated by the Great Depression, I was a week late for freshman year registration. The Dean, Dr. Charles Wright, helped me register. He insisted that I take a chemistry course which would qualify me to take more chemistry. By the end of my freshman year, encouraged by the only chemistry professor, Dr. Frederick C. Kirchner, I announced that I was majoring in chemistry.

My holistic experience among the smells of both organic and inorganic molecules in the basement of Memorial Hall gave me a special feel for the behavior of molecules. Our view was simple. We trusted our noses. "If you can't smell it, it won't hurt you", we told ourselves.

The aromas of the old chemistry laboratory were almost entirely missing from the science building where I now lectured. (OSHA regulations had finally reached the campus.) Even so, the college was abuzz over prospects for a new science building to be erected at the head of the campus.

Strangely, Dr. Kirchner often got my attention, especially in qualitative analysis laboratory, by saying, "Why Mr. Crummett, not all is gold that glitters!" I was often reminded of this during the research on trace quantities of dioxins in the environment. The false urge to always report positive results plagued all investigators, me included.

I left Bridgewater College in 1943, believing in molecules in the physical world. I still do.

THE LAW

"Even when laws have been written down, they ought not
always to remain unaltered",
Aristotle, c.384-c.311 BC.

1995

It was the best of laws. It became the worst of laws. It was and is
the "Delaney Clause", an amendment to the 1958 Federal Food Safety
Law. The Delaney Clause reads: "no additive shall be deemed to be
safe if it is found to induce cancer when ingested by man or animal,
or if it is found, after tests which are appropriate for the evaluation for
the safety of food additives, to induce cancer in man or animals". At
the time of its enactment it was sorely needed and served a great
purpose.

Scientific progress has now shown the Delaney Clause to be too
fear provoking and restrictive as it is applied today. This is due to the
development of more sensitive measurements. First, toxicologists have
bred rats and mice that are much more sensitive to carcinogens. Second, analytical chemistry methods have gone from the part per million range to the parts per trillion. Thus they are more sensitive by 4
to 6 orders of magnitude (10,000 – 1,000,000). Finally, we now
know that man-made chemicals are not the primary cause of cancer.
We also know that carcinogens abound in nature and that the human
body generally handles them quite well.

It is sad but the Delaney Clause truly is a scientific anachronism.
Some might say it is outmoded. Using modern analytical methods
for measuring detectable quantities, its requirements can never be
met.

A concentration unit is needed to assure folks in the EPA and

FDA that the zero tolerance limit for carcinogens, according to the Delaney Clause, is met. It has been suggested that the new unit proposed for use by geochemists be adopted for this purpose. The unit is called the nonogram and has the magnitude of $10^{-\infty}$.

References

1. H. E. Stokinger, "Sanity in Research and Evaluation of Environmental Health", Science, November 12, 1971.
2. Elizabeth Whelan, "Delaney Clause Should Be Repealed", Human Events, August 8, 1992.
3. David Hanson, "Flaws in Delaney Clause Make Change Imperative, But Congress May Balk", C & E News, September 12, 1994.
4. Geotimes, November issue, 1970.

RULE OF REASON

*"That one hundred and fifty lawyers should do business together
ought not to be expected",
Thomas Jefferson.*

1977

Before "Christmas Interruptus" lawyers were not a noticeable part of my life as an analytical chemist. Those who questioned my data and conclusions did so with appropriate respect and an open mind. Discussions always reached mutually agreeable conclusions. New perspectives came out of such dialogues. Patent attorneys, with whom I sometimes worked, almost always added something to the meaning of data by providing a different perspective. Of course I knew other Dow lawyers, but only in contacts outside Dow. For example, I met Dow's corporate counsel while serving on the committee to form Delta College.

Now Dow's legal department seemed to double in size overnight. Lawyers from outside law firms visited Dow regularly. Lawyers representing EPA, USDA, FDA, and NIOSH appeared at scientific conferences and, those few friendly to industry, visited Dow. All of them seemed to want to understand the meaning of analytical data near the limit of detection. We tried to explain it to them. Generally they were not happy with concepts such as "confidence levels" or "uncertainty". They were seeking firm "yes" or "no" answers to their carefully crafted questions.

Dow now sought outside counsel on all matters subject to litigation. One trial attorney was especially outstanding in using his experience to develop a "rule of reason" approach to the resolution of scientific and technical disputes with the aim of lowering the cost and raising the quality of litigation. Milton Wessel thus provided the motivation and guidance for the conduct of scientific conferences and workshops in which scientists, saying entirely different things about the same scientific question would meet together, review all the data, question each other, and reach consensus. Two such conferences were held: March 8-9, 1973, a "Rule Of Reason Workshop" in Washington, D.C., and June 4-5, 1979, a "Dispute Resolution Conference on 2,4,5-T" in Arlington, VA. Although scientists from different disciplines and persuasions appeared to reach consensus, soon newspaper articles reported the same sensational and unreasonable quotes from participants as before.

So, unfortunately, lawyers continue to be a major force in the scientific debate. Those involved in environmental matters related to chlorinated hydrocarbons often consult with me. So far I have attempted to educate more than 200 lawyers in matters related to the behavior of trace amounts of chemicals in both living and inert systems.

At the same time lawyers, together with the public relations people, have infiltrated the scientific community to such a degree that scientific considerations are no longer the driving force in conducting experiments. Rather the question of legal vulnerability and possible resulting headlines are dominant considerations. Thus the freedom of research, long enjoyed by scientists in the United States, is placed in jeopardy.

Lawyers are intelligent and logical. But their approach is in opposition to scientists. They assume a hypothesis, collect data to support that hypothesis, ignore or discredit other data, and do whatever they can to defame the opposition. Scientists, on the other hand, generate data, compare it to other data, evaluate the integrity of all relevant data, consider the preponderance of data, and develop a hypothesis. So the adversarial approach of lawyers does not seem appropriate for the resolution of scientific and technical matters unless all participants in the process are trained scientifically.

I know this because, from 1970 through 1990, I was often asked to brief attorneys on the "dioxin" question. They were always looking for authoritative answers. "Yes" or "No"! "It is" or "It isn't"! Since scientific results are seldom absolute, their questions were difficult to handle.

I am therefore, persuaded that matters of science and technology should be settled by applying the rule of reason in dispute resolution conferences as described by Milton R. Wesel, an attorney of the highest integrity.

References

1. Wesel, Milton R., "Rule of Reason", Addison—Wesley, 1976.
2. Wesel, Milton R., "Science and Conscience", Columbia University Press, 1980.

CAREERS

"The correct advice to give is the advice that is desired", "The desired advice is revealed by the structure of the hierarchy, not by the structure of technology", Archibald Putt, 1976.

1995

Although most Dow managers took measures to keep from becoming involved in dioxin issues, those who allowed themselves to suffer for a few years were rewarded with high level jobs. Apparently they had given upper management the advice they desired. Could that explain why the leader of our Dow Dioxin Initiative Team operated on the concept of "Read my lips!"? In other words, "Do as I say, but I didn't tell you!" Previous dioxin managers had encouraged scientists to present facts and make decisions based solely on the scientist's interpretation. Now I felt that I was supposed to supply data to fill in the blanks in a story already written.

After the dioxin teams were disbanded, members who were attorneys, public relations persons, secretaries, production engineers, industrial hygienists, etc., were usually given new, broader, and more interesting assignments, sometimes requiring relocation and usually included a promotion. Scientists, however, returned to their job of keeping current with the scientific literature and designing experiments to challenge the hypotheses already announced. New research on the dioxin issue now became very difficult to fund and so many research ideas were never explored.

Spontaneous research was now dead. Scientists needed to have a compelling reason, which met the approval of a battery of managers before any research relating to "toxic" chemicals and the environment could be funded. Compliance and perception became the goal of the company – not scientific leadership. "We don't want to be on point of any environmental issue", we were repeatedly told. I felt that the environmental extremists and government regulators had won. Scientists and society at large had lost. Short term, of course, the company saved resources. Long term employees and stockholders would likely pay a dear price.

I did not fully understand this change in philosophy until 1995, when I first learned of Dow's "The Futures Group" as described by Patrick P. McCurdy, who wrote, "For Dow Chemical, 1983 was the 'year of the dioxin' in which the company, during the early spring and for weeks, was pilloried (unfairly, in my view), literally every day by the press and electronic media over the dioxin issue, especially as it related to Agent Orange. You might have thought that Dow, given its previous experience over the napalm questions, would have been well prepared to handle this public relations issue (which was largely orchestrated by lawyers for Vietnam veterans claiming injury, feeding a naïve and gullible press).

"This was a crisis unlike the others. There had been no plant disaster, no deaths. But treatment of the company by the press became a sort of "Chinese water torture" of calumnies. Although Dow handled the day-to-day media queries well enough, even after a month or so, the thing, like a bad cold, just wasn't going away, and the company seemed to be slowly sinking into a PR quicksand not of its own making. Finally, the company got a grip on itself, appointed a PR point person, hired a PR firm to do fire fighting and, most importantly, funded a multifaceted investigation of alleged problems involving dioxin and then invited fifty-plus media representatives to Midland, MI, to a press conference, where the company's top brass, including CEO Paul Oreffice, outlined the steps that were to be taken."

"Press hits, which had been rising, fell off the cliff. Even more important, company management decided to study its corporate navel and set up an internal task force called "The Futures Group". The group's mission: Find out if the company had a chip-on-shoulder image problem (it did) and, if so, what to do about it. One result, among many, was the inauguration of a new corporate advertising campaign focused on the theme, "Dow Lets You Do Great Things".

The campaign has been running close to ten years now and, I am told, has achieved good results. (Incidentally, you might wonder how I know all these things. Simple, I worked for Dow, 1980-84, and was part of that Futures Group.)"

The campaign included a stable of 19 Dow spokespersons who gave 379 interviews and editorial briefings during an 18-week media tour. I was invited to join the stable but declined.

Reference

1. P. P. McCurdy, "Dow Chemical and the year of the dioxin (1983). "PR Battles Lost and Won", Today's Chemist At Work, February 1995, p 96.

DIOXIN IN ME

"A thousand probabilities do not make one fact",
Italian proverb.

2000

I would like to know the 2,3,7,8-tetrachlorodibenzo-p-dioxin content of my own body fat. I postulate that it would be at least an order of magnitude higher than the average of 6 parts per trillion generally found in the adipose tissue of American citizens.

Such data would give me an excuse to exhibit righteous indignation: at society for being so ignorant, at government agencies for not protecting me, at schools for not informing me, and chemical companies for not being clairvoyant and designing appropriate safety rules during the early part of my career.

Such data might explain my physical symptoms: continuous severe ringing in my ears, tingling buzzing in various parts of my body, sore aching joints and muscles, plugged sinuses, stiff neck, dry eyes, and too frequent urge to urinate. Physicians say I am in good health. They say nothing will help these symptoms. They say the symptoms are only a part of the aging process. At age 80, I find little research being done to correct these annoyances and distractions.

I was born and raised in houses heated by wood-fired iron stoves and fireplaces lighted by tallow candles and kerosene lamps. All of these devices emitted traces of smoke into the house on a continuous basis. On being stoked, poked, or adjusted they belched much larger quantities. These combustion activities alone were sufficient to assure my constant exposure to trace products of combustion. In addition, much food was fried in lard and often charred in the process. And then there were the outdoor activities – open wood fires under kettles of hot water for the weekly clothes wash, apple butter boiling, maple syrup concentration, soap making, etc., almost all the time. All of these produced enough smoke exposure to cause the eyes to burn, tears to flow, and lungs to cough. Fortunately my family did not smoke tobacco but most persons felt free to smoke when they visited us and our 90-year old neighbor smoked a pipe constantly. In addition, there was a frequent haze from forest and brush fires. These were the experiences of the first 17 years of my life.

In college I shared an apartment with four other boys. We stoked a wood fired stove and cooked our meals on an oil range with an open flame. The coal-fired college boiler spewed black smoke over the campus and surrounding area. Leaves were raked off the campus lawn and burned in the street. Faculty and students stood around and enjoyed the smell of burning leaves. The chemistry laboratory, in which I spent a lot of time, had no hoods, but every student had a gas fired Bunsen burner burning most of the time. Tobacco and herb smoking were not allowed on campus, but students and workers smoked in the street. Thus I had another four years of exposure.

As a control chemist in the chlorine plant of the Solvay Process Company, I undoubtedly was exposed to a large variety of chlorinated organic compounds. All production plants leaked to some degree. The powerhouse rained black smoke and soot down on us. Powerhouses of at least two other large companies added to the high level of smoke. Tobacco smoking was permitted in the plant and the laboratory as well as in all local business establishments. Thus is added another three years of exposure.

In graduate school in Columbus, Ohio, I walked about four blocks from my residence to the chemistry department. On arrival I often found my clothes to be smudged with black soot. On such occasions I usually blew such soot from my nostrils. This soot came from the burning of soft coal, which produced enough soot to cover almost everything. As before most men smoked tobacco in the laboratories and anywhere else they pleased. When the validity of the "trace chemistries of fire" became evident I found myself thinking very hard about these exposures.

By 1951, I had chosen to work for The Dow Chemical Company

in Midland, Michigan, because the air appeared clear and odors were less than at most other industrial sites. Although our drapes did not become black with soot as they had in Columbus, fly ash was plentiful and I had cinders removed from my eyes about twice each year. More ominous, laboratory benches would quickly become coated with dust. (Twenty-five years later similar dust was found to contain trace amounts of most chlorinated dioxins.) Meanwhile, in America, the number of automobiles and diesel trucks was increasing dramatically. All belched fumes. Around Dow a faint odor of chlorinated phenols was often noticeable. Inside the production plants the intensity was almost overpowering. In the laboratory I often worked with aqueous solutions of the salts of these phenols. Sometimes, in haste, I spilled small quantities of these on my skin. And there was always the open flamed Bunsen burner at each workstation. We synthesized 2,3,7,8-tetrachloro-p-dioxin in our laboratory without any health incidents. We developed analytical methods and collected many samples suspected of being contaminated with chlorinated dioxins. Standard samples of these dioxins were obtained and stored. We worked very carefully and no one got chloracne. But, even so, a few molecules would surely have gotten into our physiological systems. This continued for another 24 years. Meanwhile, I was required to attend many, many meetings where most attendees smoked with great gusto. Wood was burned in fireplaces in our neighborhood and vehicular traffic increased. Lawn mowers belched more smoke. More chemicals were needed to keep lawns neat and green.

By all extrapolations of toxicology data by soothsayers such as risk assessors and environmentalists, I should have died a horrible death at least ten years ago. So I feel pretty good that I can still walk 18 holes of golf, and swing a club a large number of strokes.

So why don't I know my dioxin content? Certainly the Wizards would delight in sizzling my fat and adding the dioxin results to their swelling body of classical data. Perhaps my personal "contamination" should be a matter for public record. Acquiring such information, in my judgement, would not justify surgical invasion to take a sample: I am not persuaded that levels of dioxins in human fat at

100 parts per trillion is a risk greater than the surgical intrusion necessary to take a sample. So I keep my dioxin intact.

Perhaps I should calculate the amount of dioxin in my fat. First, I can estimate the amount of exposure to dioxin I have had from smoke from wood fires (wood stoves, fireplaces, open kettle fires, brush burning, leaf burning, forest fires, and bon fires). I can add to that the amount from the burning of hydrocarbons (kerosene lamps, gasoline lamps, kerosene stoves, candles, Bunsen burners, automobile exhaust, and diesel exhaust). Then I can add exposure from waste incinerators (municipal, industrial, and hospital). Other things to add include emissions from coal fired powerhouses and the degradation products from rubber tires on asphalt roads. Finally, I can add exposure to a disinfectant (hexachlorophene), brush killer spray, the atmosphere of chemical laboratories, and leaks from chlorine production plants.

Second, I will make my own estimation of the percent of dioxin getting into my fat from each type of exposure.

Third, I will postulate the rate of decomposition of dioxin in my body fat along with the rate of elimination.

Fourth, I will calculate my body content due to each six-year period of exposure using a computer program capable of producing at least four significant figures.

Finally, I will do a summation of the resulting 18 numbers like environmentalists do. The result will be publishable if I am able to explicate each step in the appropriate jargon. Doing this, however, makes me feel scientifically guilty. Unlike some, I don't understand what an estimate times an estimate times a postulate really means. I think your guess is as good as mine.

Second by second, minute by minute, hour by hour, day by day, week by week, month by month, year by year, time accumulates until at my age more than 3.3 billion seconds are registered. I calculate that, assuming my fat contains 60 parts per trillion of 2378-TCDD and no degradation has occurred, I would have accumulated it at an average minimum rate of 54,000 molecules per second. The average person (my age and weight) having 6 parts per trillion would have had a rate of 5,400 molecules per second. If all chlorinated compounds were totally banned by governments, the rate would never get lower than 1,000 molecules per second.

This means that I probably now have 26,000 molecules of TCDD per cell. Of course, these are really not evenly distributed. Most would be in my adipose tissue, and not easily available to the rest of my body.

Reference

1. H. Poiger and C. Schletter, "Pharmacokinetics of 2,3,7,8-TCDD in Man", Chemosphere, *15*, Nos. 9-12, p. 189 (1986).

THE PAY OFF

"Nothing in life is to be feared. It is only to be understood!"
Marie Curie,
Polish Nobel Prize Winning Chemist, 1867-1934.

The strategy for the management of dioxin molecules was simple – remove them totally from the environment. Many regulatory limits have been put in place, and industry and government have spent many billions of dollars to accomplish this. However, this has been and remains a vain goal, because the nature of dioxins is such that their concentrations can never be reduced to zero. Thus this enormous effort, insofar as dioxin levels are concerned, has had no measurable effect on the life expectancy of the people of the world nor on their daily well-being. Likewise the health of ecological systems has not been improved by the attempt to remove dioxins from the environment. Therefore, it is reasonable to believe that the many billions of dollars spent could have been used to manage molecules and situations having clearly proven high risk factors. Thus a major scientific conference[1] concluded that, "2,3,7,8-TCDD is not a chemical rating an exceptionally high public policy priority which diverts resources and public attention from other widespread and more dangerous chemical compounds".

These observations suggest that the great effort on dioxins has been wasted. However, that is not the case. The research effort on dioxins has produced several unexpected benefits. From a scientific viewpoint, these include the following:

1. great advances in the analytical chemistry of trace quantities,
2. great advances in toxicology,
3. the development of models for risk assessment,

4. more thorough epidemiological studies
5. proof that chemophobia is a severe malady, and
6. better understanding of how scientists should report to the
   news media.

As a result of the dioxin effort, analytical science has advanced rapidly in sensitivity, selectivity and reliability. At Dow, the limit of detection was lowered in a step-wise fashion from 1,000,000 parts per trillion in 1965, to 0.01 parts per trillion in 1983, to 0.001 parts per trillion in 1985. In 1966, the number of possible tetrachlorodibenzo-p-dioxin isomers included in the 2,3,7,8-tetrachlorodibenzo-p-dioxin chromatographic peak was 20; and in 1979 only one, the 2,3,7,8-isomer alone. This made the method specific for 2,3,7,8-TCDD and together with other concurrent developments, minimized the number of possible false positive and false negative results. The practical result of all this effort, however, is that measurements made at higher levels, such as parts per billion, can now be made with complete confidence.

For many years animal toxicology accurately predicted the acute toxicity of chemical compounds in humans. Dioxin, more than any other molecule, required that chronic toxicity be predicted. In the effort to achieve this, the importance of several very fundamental experimental parameters was necessarily re-emphasized and reduced to common practice. Among these were: the identity and purity of the compound being tested; the method of introduction of the compound into the animal; the difference between malignant and non-malignant lesions; the measurement of the amount of the compound absorbed, accumulated, and excreted; and models for the extrapolation of animal data to humans at exposure levels lower than those to which the animals had been exposed. The difference in the toxicity of 2,3,7,8-TCDD to different animal species, although a phenomena previously observed for other compounds, was so pronounced that extrapolation of its toxicity to humans required special investigation and data treatment. In the process of doing this, many new approaches and techniques were developed. This contributed greatly to the evolution of the field of toxicology. Certainty in the meaning of data has been markedly improved from the state of the field in 1972.

In the early 1970s, risk assessment was based on an educated guess. Calculations were made with numerous safety factors included. Thus, risks from chronic exposure were almost always estimated to be very high. Over the years many models for assessing risk have been proposed. A few of these have emerged as reliable.

Although appropriate protocols for assessing risk through epidemiological studies have been available, many such studies have been done with inadequate funding, or with inadequate safeguards to avoid the influence of compounding factors which bias the results. The many thorough reviews of such studies by scientists, lawyers, and the public in the dioxin cases, has resulted in government and industry studies being much more alert to other possible contributing toxicity factors which must be controlled or accounted for in the study.

The dioxin experience has proven that chemophobia is a serious psychological affliction of many people. In the beginning, the overreaction to the existence of dioxin was understandable. Billed as the "most toxic man-made" chemical – it scared everyone. Although experienced in working with hazardous chemicals, I was frightened to have my co-workers handling this material. Soon, however, I learned that this material, like all other material, consists of molecules, which can be managed and kept from harming anyone. As scientific evidence accumulated, much of the anecdotal information was proven unreliable and fear of traces of dioxin in the environment was greatly reduced among scientists. Some other interested parties failed to accept the new information and continued to insist that "the sky was falling". Their failure to accommodate new scientific evidence to allay fears but twist it to enhance fears proved they are victims of chemophobia. The epidemic of chemophobia is a major problem for a modern society. All society working in concert to treat those afflicted with this malady will be needed to correct the problem.

Because the trace chemistries of fire hypothesis showed that dioxins are produced during incomplete combustion, great advances have been made to assure that combustion is complete in municipal and industrial incinerators, powerhouses, and wood stoves. Although the reduction in dioxin content cannot be shown to have a positive impact on human health, the process virtually eliminates other noxious materials in the smoke, resulting in a much healthier environment.

To eliminate dioxins from air, water, and soil, new methods of handling and cleaning these materials have been devised which virtually eliminates other toxic materials. While the reduction in dioxin is insignificant from a human health prospective, reduction in other toxic materials makes the environment a more pleasant place.

In perspective, the extraordinary effort made to eliminate dioxins from the Midland, MI, environment has not created a safer environment from the reduction in the concentrations of dioxin. But other possibly toxic materials originally present at concentrations of a million or more times higher than the dioxins have been greatly reduced.

References

1.M. A. Kamrin and P. W. Rogers, "Dioxin in the Environment", Hemisphere Publishing Corp., New York, 1985, p 280.
2.H. Hillman, "Certainty and Uncertainty in Biochemical Techniques", Ann Arbor Science, Ann Arbor, MI, 1972.

LESSONS, CONCLUSIONS AND PREDICTIONS

"In all things there is order, harmony, and wisdom",
H. Davy, 1811.

2000

Scientists are set on a course to detect and measure molecules at lower and lower concentration levels. The trend is inexorable and will only stop when single molecules can be seen, identified and counted one by one. As molecules are identified and measured at lower levels, the number of different "hazardous" molecules found will increase exponentially. Over the past 20 years, each new discovery of dioxin, regardless of concentration, has resulted in instant experts and reporters producing scary stories. Simultaneously, predators (special interest groups) start circling the institutions with whom the dioxin is associated. As dioxin detection goes to lower and lower levels, no company engaged in any sort of commercial or industrial activity will be exempt. This will create an unprecedented challenge to toxicologists and risk assessment experts to provide data on which to base sound decisions and to allay newly created public fears.

These public fears will be created by special interest groups who will continue to demand that "zero" molecules of any synthesized toxic substance be allowed in the environment. If scientists continue to behave like those who practice philosophism, setting themselves up as instant experts and making sensational statements to the media, the brouhaha will intensify.

The situation is worsened by the disdain academic and government scientists often show to scientists in industry. In my experience, there is no difference in scientists regardless of where they work.

They are all human and therefore, sometimes fallible. They all have vested interests. Yet, they share a common purpose and are generally honest in depending on data and tested hypotheses for decision making. It should be recognized, however, that industrial scientists enjoy using the finest facilities available. Furthermore, they have also been educated at the best universities. So data collected by them are usually the best that can be obtained.

The perception still exists that, "We can't believe him. He's an industrial chemist". The implication is that industrial chemists are told what to say by management. It is difficult for an industrial scientist to understand how this perception arose. Perhaps it has resulted because industry uses "spokespersons" to interact with the media. These persons put into simple language what they believe to be the truth about their environment. They are not scientists, although they sometimes speak for scientists. Somehow, scientists are viewed as naïve. They are thought to say things not in their own best interest. This perception needs to be changed, and the media and industrial scientist should work together. It is generally easy to identify the true scientists. Just ask them how many measurements they have made themselves.

More importantly, academia, government, industry, environmental groups, and the media should work together. I have been unable to detect much difference in the goals of these groups. All want to live in a natural environment. All want to benefit from the modern conveniences that technology has brought. All want to learn more about life and its meaning. The approaches are different and the language is different. For any one of these groups to set itself up as the final authority is counter-productive and is a disservice to society as a whole. There needs to be better communication among all five groups with the goal of educating the public in a more realistic sense. If such systems are to be achieved, interaction among the scientists themselves in the various disciplines must be greatly enhanced. This should speed up the debate and permit consensus to be reached more quickly. Once scientists are doing this complete job, information given to decision-makers in all groups will be consistent, resulting in better decision making.

There is no doubt that dioxins are molecules (not evil spirits) and can, therefore, be managed. So all chemicals can be managed. Society has thus far chosen to do this through regulation. While this approach has brought substantial benefits, it also has its problems. The most dangerous of these is the effect on the scientific process, especially in industry. Politics is now driving science. When the greatest advances in technology were made 1950s and 1960s) science was driving politics. A proposed scientific experiment in industry today must have an answer to the questions: What affect will this have on public relations and the law? Usually, at least one decision-maker can find a reason why doing the experiment is not in the best interest of the company. Simply put, scientists are not as free to study the problems they see around them as they once were or should be. We need to ask the question: Thus reined, will industrial scientists still be able to recognize a potential future hazard and develop means for managing it as they did for dioxin? Unfortunately, the answer is: not likely.

Yes, chemicals can be managed. Spills, leaks, and air emissions are now easily detected at concentration levels far below those of any health concerns. More importantly, systems are in place to prevent these from happening. But when a ubiquitous chemical such as TCDD is attempted to be managed at part per quadrillion levels, a concentration probably below its normal concentration in drinking water, the problem becomes extraordinarily difficult, if not impossible. In this case, the view that dioxin is more than a molecule appears to be in the minds of the decision-makers.

Concerns about dioxins at barely detectable concentration levels are frequently expressed in the newspapers. However, everyday misuse and mismanagement of chemicals having 100,000 to 1,000,000 times greater risk than dioxins are ignored. These abuses include: the black smoke that belches from diesel trucks when they change gears; the fumes released when trucks are left idling, bombs are exploded, or canons are fired; uncovered trucks driving at high speeds with dust flying from their beds; smoke from home heating units; smoke from forest fires; smoke from burning landfills; rubber and asphalt particles on highways; gasoline fumes at the gas pump, etc. The list is quite long. Yet environmental groups and newspapers fight over parts per quadrillion levels of dioxin in river water and part per trillion levels of dioxin in paper products, while accepting risks of a chemical nature in these everyday happenings. In their view, dioxin must be an evil spirit

Concerns about dioxins at concentration levels so very, very low raises the question: Should we be concerned about trace levels of "naturally occurring" toxic molecules in foods? Almost all foods have been shown to have some. Some foods taste great, but we don't feel great later. However, if we manage ourselves and eat only small portions, everything is fine. This is an experience all of us have had, and an experiment we all can do.

The lessons learned from the dioxin issue are vital to each individual in the world. And each individual has a stake in seeing that molecules are managed properly. To do this requires interaction with persons in the chemical industry and other related industries, remembering that molecules must be counted before they can be managed. Visit the facility, invite representatives (especially scientists) to give talks to your clubs and organizations, and be prepared to share your perception. Remember that in the final analysis industry people are working for you, the consumer. But they need your point of view. If laymen continue to be aloft, dioxin and other molecules will indeed become evil images.

Unfortunately, there is no end to the dioxin story. Dioxins are still with us and will continue to be discovered wherever carbon atoms and chlorine atoms are heated together. Chlorinated dioxins will be formed at levels below parts per billion. Such levels should present no health problems to those who believe that dioxins are molecules. But to those who don't they are indeed evil spirits.

THE SCIENTIFIC COMMUNITY

"Prove all things; hold fast that which is good", New Testament,
I Thesselonians.

The scientific community continually convenes symposia and workshops devoted to a review of the scientific data and an evaluation of the state-of-the-science. Many such conferences have addressed scientific questions related to the chlorinated dioxins and related compounds. These conferences have been well attended by the most eminent scientists in the various fields of science related to the chlorinated dioxin issues. I have been privileged to participate in the following such conference/workshop meetings:

* "Symposium on Trace Analysis", The New York Academy of Medicine, New York, NY, November 2-4, 1955.
* International Symposium on "Identification and Measurement of Environmental Pollutants", Ottawa, Canada, June 14-16, 1971.
* "Dioxins – Origin and Fate", American Chemical Society, Washington, D.C., September 16-17, 1971.
* "Rule of Reason Workshop on 2,4,5-T", Washington, D.C., U.S.A., March 8-9, 1973.
* "Conference on Chlorinated Dibenzodioxins and Dibenzofurans", National Institute of Environmental Health Sciences, Research Triangle Park, N.C., U.S.A., April 2-3, 1973.
* "Health Effects of Halogenated Aromatic Hydrocarbons", The New York Academy of Sciences, New York, NY, June 24-27, 1978.
* "Examination of the Scientific Basis for Government Regula-

tions", Engineering Foundation Conference, Franklin Pierce College, Rindge, NH, July 9-14, 1978.

* Ninth Annual Symposium on "The Analytical Chemistry of Pollutants", USEPA, ACS, University of Georgia, Jekyll Island, Georgia, U.S.A., May 7-9, 1979.

* Dispute Resolution Conference on 2,4,5-T. American Farm Bureau Federation, Arlington, VA, June 4-5, 1979.

* Collaborative International Pesticides Analytical Council (CIPAC), Twenty-third Meeting and Symposium, Baltimore, Maryland, June 6-7, 1979.

* Chlorinated Dioxins and Related Compounds. Impact on the Environment", First International Symposium, Instituto Superiore Di Sanita, Rome, Italy, October 22-24, 1980.

* "Harmonization of Collaborative Analytical Studies", International Union of Pure and Applied Chemistry, Helsinki, Finland, August 20-21, 1981.

* "Human and Environmental Risks of Chlorinated Dioxins and Related Compounds", Second International Symposium, Arlington, VA, U.S.A., October 25-29, 1981.

* "Pesticide Chemistry – Human Welfare and the Environment", International Union of Pure and Applied Chemistry, Kyoto, Japan, August 29 – September 4, 1982.

* "Chlorinated Dioxins and Related Compounds 1982", Third International Symposium, Salzburg, Austria, October 12-15, 1982.

* "Public Health Risks of the Dioxins", The Rockefeller University, New York, NY, U.S.A., October 19-20, 1983.

* "Dioxins in the Environment", Center for Environmental Toxicology, Michigan State University, East Lansing, MI, U.S.A., December 8-9, 1983.

* "Environmental Standards Workshop", Argonne National Laboratory and the American Society of Mechanical Engineers, McLean, Virginia, U.S.A., January 23-26, 1984.

* "Chlorinated Dioxins and Related Compounds", Fourth International Symposium, Ottawa, Canada, October 16-18, 1984.

Don Townsend and I finally reach Salzburg, Austria.

* "Dioxin '85. 5th International Symposium on Chlorinated Dioxins and Related Compounds", Bayreuth, FRG., September 16-19, 1985.
* "Dioxin '86. 6th International Symposium on Chlorinated Dioxins and Related Compounds", Fukuoka, Japan, September 16-19, 1986.
* "Dioxin '87. 7th International Symposium on Chlorinated Dioxins and Related Compounds", Las Vegas, Nevada, U.S.A., October 4-9, 1987.
* "Dioxin '89. 9th International Symposium on Chlorinated Dioxins and Related Compounds", Toronto, Canada, September 17-22, 1989.
* "Dioxin '90. 10th International Symposium on Chlorinated Dioxins and Related Compounds", Bayreuth, FRG., September 10-14, 1990.
* "Dioxin '91. 11th International Symposium on Chlorinated Dioxins and Related Compounds", Research Triangle Park, North Carolina, U.S.A., September 23-17, 1991.

Each of these conferences summarized the work that had been done and evaluated its integrity. Points of agreement and disagreement among scientists were generally clearly defined. Moreover, recommendations for further research were made in a clear and concise manner. Any agency or institution desiring to take responsible action or promote useful research needed only to examine the proceedings of these conferences. To my knowledge, this is not systematically done. Those involved invariably choose to create new issues or, failing that, to seize upon the issues generated by anecdotal evidence (old wives tales, suspicion, fear, etc.). The ignoring of scientific evidence together with the elevation of anecdotal evidence makes it very difficult to solve perceived problems and reminds us that the human race has not yet completely extracted itself from the bondage of superstition.

On route to Dioxin '91 on Interstate 40 in Tennessee, we were welcomed by a gigantic billboard altering the mountain scenery.

Many of the conferences resulted in books being written. Selected comments made by the editors are interesting and useful in making judgments about action concerning dioxins.

1973   J. A Moore, " . . . . to provide a forum where the current knowledge of these compounds could be presented and licensed by the government, industry, and university-community."

1978   F. Cattabeni, A. Cavalloro, and G. Galli, "It seemed there fore, necessary to hear from the scientists regarding the recent advances on the chemistry, analysis, toxicology and decontamination of TCDD".

1982   O. Hutzinger, R. W. Frei, E. Meriam, and F. Pocchiari, "The complex, multi-disciplinary problem poised by PCDD can only be understood by interaction of scientists from different disciplines . . .".

1982   O. Hutzinger, R. W. Frei, E. Meriam, and G. Reggini, "This topic still commands considerable attention from various points of view and the successful symposium has once again seen participation of many of the leading scientists who are active in this area". 1983.

1983   R. E. Tucker, A. L. Young, and A. P. Gray, "Not only did scientists representing the various disciplines covered in the program attend, but also lawyers, engineers, economists, and representatives of the media. It is especially interesting to note that over 70 physicians attended the four-day symposium.

1983   G. Choudhary, L. H. Keith, and C. Rappe, "The 1982 symposia brought together experts, both national and international, to a common forum".

1983   W. W. Lowrance, "This symposia was organized as a meeting of scientists and physicians to discuss the difficult technical issues, not to argue about political, legal, economic and other nonscientific issues, which need different forums. Participants included a wide variety of scientific experts and leaders from industry, government, special-interest groups, academia, and the press".

1983   M. A. Kamrin and P. W. Rodgers, "These papers were presented before nearly 300 scientists, decision makers, and concerned citizens who were in attendance.

1983   M. J. Boddington, P. Barrette, D. Grant, R. J. Norstrom, J. J. Ryan, and L. Whitby, "This topic still commands considerable attention from various points of view and the successful symposium has once again seen participation of many of the leading scientists who are active in this area."

1983   Y. Masuda, O. Hutzinger, F. W. Karasek, I. Nagayama, C. Rappe, S. Safe, and H. Yoshimura, "Each symposium in the series contributed excellent papers, especially in the 5th International Symposium in West Germany, many scientists presented remarkable results of their research."

1987   D. N. NcNelis, C. H. Nauman, L. K. Fenstermaker, S. Safe, R. E. Clement, J. Campan, P. Kahn, M. Fingerhut, P. desRosieres, E. J. Sampson, R.K. Mitchum, C. Rappe, D. G. Barnes, F. Hileman, and L. H. Keith. "The outstanding number of excellent papers and overwhelming attendance at the Symposium clearly indicates the increasing awareness and interest in the environmental and human issues pertaining to Dioxin and its related compounds.

1989   R. Clement, C. Taskiro, G. Hunt, L. LaFleur, and V. Ozvacic, "The symposium—was attended by over 700 delegates. . . . about 300 presentations were given . . . . The number of papers show that the field of chlorinated dibenzo-p-dioxin and dibenzofuran research is still extremely active, and that significant scientific issues remain to be solved in spite of many years of active investigation."

1991   T. Danstra, L. Birnbaum, M. J. Charles, C. Corton, W. E. Greenbe and G. Lucier. These editors made no published comment.

As more data are collected, evaluated, interpreted, and discussed, scientists eventually reach a consensus. Consensus has been reached in most aspects of the chemistry of dioxins.

The growth of the dioxin business can be seen from the growth of scientific activity as reported at international conferences.

Year	Conference Location	Number of Authors	Pages Published
1973	Research Triangle Park, NC, USA	86	313
1980	Rome, Italy	117	649
1981	Arlington, VA, USA	128	797
1982	Salzburg, Austria	129	361
1984	Ottawa, Canada	169	414
1985	Bayreuth, Germany	436	1047
1986	Fukuoka, Japan	208	607
1987	Las Vegas, NV, USA	461	1339
1989	Toronto, Canada	628	1249
1990	Bayreuth, Germany	1068	2043

An even more effective conference was held at Michigan State University, December 6-9, 1983. Scientific work on "Public Policy on Dioxins", "Human Health and Toxicity", "Source, Distribution, and Fate", and "Sampling and Analytical Techniques", were reviewed and summarized. This was followed by workshops in which a summary of knowledge was made and suggestions for future studies were advanced. Journalists were in attendance and participated in the workshops.

Even so, less than a year later I reported in Ottawa, Canada, that the dioxin issue had escalated from science to a broad social issue. So "public perception, not science, was controlling the research effort". This, I said, was due to "phenomenal gaps between the science and what the public understands". "As scientists, we had not succeeded in relating these increasingly complex subjects to the real world." I urged that we increase our efforts.

By 1987, it was evident to me from private comments by leading scientists, that the scientific effort had been successful. "Dioxin" had been found in many things and the means by which it got there was well understood. Scientific data were available to correct the most troubling situations. The science was there to solve most foreseeable problems. So, I could now retire with confidence that humans need not fear "dioxins" any longer. Any further concern would be about politics – not the scientific truth. By my last conference (1991) politics seemed pretty firmly entrenched in the scientific conference, however.

IN RETROSPECT

"Analogies prove nothing, that is true, but they can make one feel more at home",
Sigmund Freud, 1932.

There was an old lady who lived over the hill from my boyhood home. She was by far the oldest woman around, being more than 90 years old. I visited her frequently, usually with my parents. She sat in a corner of the kitchen by the large fireplace, where her collection of herbs dried on the wall. She smoked these herbs in a pipe having a homemade baked clay bowl and grapevine stem. She also brewed potions from these herbs. She had a different herb for each type of pain – boneset for backache, catnip for toothache, peppermint for stomachaches, etc. She wore many petticoats and a shawl made of homespun wool. Everyone affectionately called her "Ole Granny".

"Ole Granny" appeared to me to know everything about life and nature, but she once told me about an experience which, as a lad of seven years, I did not understand. She had been visiting her relatives who lived at the head of a hollow in the eastside of Hoover Mountain about two miles from her home. On her return, she walked across a pasture field on the western mountainside.

There she witnessed a strange sight which "skeered" her "considerable". What she at first thought was a barrel hoop rolling toward her was indeed a hoop snake. This snake formed a hoop by putting its tail in its mouth and rolling along like an automobile tire. Everything the snake touched died. Even large oak trees would tremble and within a few days drop their leaves and acorns. The hoop snake was the most poisonous and dangerous being in the world.

I could not understand how she escaped with her life and told her so. It was easy, she said. Once the snake got rolling with its tail in its mouth it could not change direction until it hit something and got stopped. So she just stepped aside. I had other questions such as how could anything be so poisonous. At that "Old Granny" responded in low German, a language which she well knew I did not understand.

Her story bothered me a lot. At first, I believed it totally for how could anyone dream up such a tale. Later I decided that "Ole Granny" was indeed one of the most imaginative persons I had ever known. Then in 1972, "The Foxfire Book" was published. And it describes hoop snakes and tells the same story. Apparently hoop snakes are part of the early folklore of the Appalachian Mountains, at least from Georgia to West Virginia. The media that spread this lore over so large an area consisted of wandering minstrels, circuit riders, peddlers, tramps, hunters, and soldiers.

The "hoop snake" is like "dioxin". Both are built largely on fantasy. The "hoop snake" is much more toxic than a rattlesnake, much more dangerous, and much more insidious. "Dioxin", as described by the modern media, is much more toxic, dangerous, and insidious than the molecule 2,3,7,8-Tetrachlorodibenzo-p-dioxin, described by scientists. The "hoop snake" and "dioxin" are figments of fearful human beings fertile imagination.

I was the stupid one – just now realizing that "Ole Granny" was preparing me to distinguish between the real and the imagined. This distinction should have been clear to me from the beginning.

References

1."The Foxfire Book", Doubleday & Company, Inc., Garden City, NY, 1972, pp 293-295.
2.Thomas Helm, "A World of Snakes", Dodd, Mead & Co., NY, 1965, pp 150-152.

AFT-WORD

"Old friends are the best."

To Whom It May Concern:

"When Warren Crummett came to work in the Analytical Laboratory of The Dow Chemical Company in 1951, I was especially pleased that he was assigned to work with me. He had done his doctoral work under Prof. William MacNevin at Ohio State University. MacNevin and I had both been students of I. M. Kolthoff at Minnesota during about the same periods, so I knew that the principles and discipline of that great teacher would have been passed along to Dr. Crummett. In this I was never disappointed. We collaborated in much of the early analytical research of the laboratory and when a time came for the efforts to be greatly expanded, it was a pleasure to assign a group for organic analysis problem-solving to him. The work continued to expand and has led to the tremendous analytical sciences facilities within the Dow Michigan Division today.

I can vouch for the authenticity of the incidents narrated by the author in this book, not only because many of them were described to me at the time by him or The Wizards, but also because of my confidence in Warren's reliability. He draws conclusions only when sufficient data are available and the signal to noise ratio is high.

As Warren's former, "boss", I always liked to have the last word as to how an idea should be expressed even though what he had written may have been perfectly adequate. Therefore, I have not been able to refrain from adding a few thoughts of my own about the dioxin brouhaha—

'Give a dog a bad name and you may as well shoot him.'

Why did 2,3,7,8-TCDD ("dioxin") acquire such an evil reputation that Dr. Crummett wonders whether it is really a chemical or a poltergeist (a noisy spirit or ghost)? In a glass bottle the compound looks like most other chemicals, and I had a little in my collection for quite a few years without noticing any unusual behavior. Dioxin has an extremely low solubility in water (not more than a few parts per billion) and would hardly seem qualified to act as a pollutant. Many other potentially toxic chemicals, chloroform, for example, is much more soluble and in fact are present in most municipal drinking waters in larger concentrations.

Dioxin's bad reputation has led regulatory agencies to recommend that fish from various lakes and rivers, even from the Great Lakes, not be eaten at all, or only in limited quantities. Yet the fish swim about in apparent good health despite their content of dioxins, PCB's, etc., probably happy that fewer of them are being eaten. Why don't the fish seem to be unhealthy?

The answer to both of the above questions may lie in the way toxicologists chose to do their experimental work, when they observed the low dioxin levels capable of causing toxic effects (in animals) quoted earlier in this book. The experimenters had to find a way of putting unusually small amounts of TCDD into the stomachs of their test animals. Thus they were led to prepare solutions with weighable amounts of dioxin (one cannot accurately weigh 0.2 microgram) and dilute them to the desired low concentration. Since they could not dissolve the dioxin in water, they had to use an organic solvent, and they chose acetone which itself is almost nontoxic internally. Then they mixed the very dilute acetone solution with corn oil in 1:1 ratio by volume and injected a desired amount of the mixture directly in the stomach of the test animal through a tube down its throat. This dosing procedure is termed "by gavage" and is designed to ensure that a definite quantity reaches the stomach, which would not be likely in normal feeding.

An objection to this procedure is based on the probability that a large proportion of acetone will carry dioxin through the intestinal wall and other membranes into cells where toxic action takes place.

Acetone is structurally very similar to dimethyl sulfoxide (DMSO):

$$CH_3\text{-}\underset{\underset{O}{\|}}{C}\text{-}CH_3 \qquad\qquad CH_3\text{-}\underset{\underset{O}{\|}}{S}\text{-}CH_3$$

Acetone   Dimethyl sulfoxide

DMSO is noted for a capability of carrying dissolved material through the skin, and it has been used medically as a solvent enabling remedial compounds to penetrate to subcutaneous sites. Such use has been discouraged since it was found that the solvent sometimes carries foreign material, including bacteria, through the skin as well.

This penetrating quality of DMSO is shared, though perhaps to a lesser extent, by acetone, alcohol, and probably other small molecules that are readily soluble in water. On this account, I feel that acetone should not have been used as the solvent in toxicological experiments. An alternative procedure would be to dissolve the compound in a more volatile solvent such as methylene chloride, place a smaller measured volume of this in the corn oil, allow the solvent to evaporate in vacuum or by gentle heating, and then proceed with the injection.

If my hypothesis is correct, dioxin should not appear nearly as toxic by the alternative procedure. Lacking acetone to carry dioxin through the intestinal wall, most of the compound will remain in the intestine until passed out with the feces. This experiment should conform much more closely to what would actually occur when an animal or person eats a food possibly containing a trace of dioxin. Any that is absorbed will probably be stored in fat rather than penetrating critical cells. Only a heavy drinker (of alcohol) would be likely to encounter toxic effects.

Again, if the hypothesis is correct, it may provide a clue as to why dioxin appears to have such different LD50 responses in different species. The effect could be due to varying permeabilities of the membranes to acetone. This possibility can be investigated by permeability studies on intestinal walls for the various animals, using techniques already developed for testing plastic films. Extension to membranes from human victims of accidental death should yield results that would allow much better risk assessment based on animal data.

The poltergeist quality of dioxin seems to go along with the frequently repeated opinion that "TCDD is the most poisonous chemical known to man". While I have my doubts about this, many people, scientists among them, accept it as gospel. Dr. Samuel Epstein is reported to have said that TCDD is 80,000 times more toxic than cyanide, presumably basing his statement on the lowest LD50 data. I would like to make a standing offer to Dr. Epstein: if he will eat or drink 80,000 micrograms (0.080 gram) of potassium cyanide, in any form he wishes, I will eat one microgram of TCDD in a bowl of cereal. This is how confident I am about the lack of dioxin toxicity when not injected with acetone. However, he had better have an antidote handy if he tries the test.

If further research confirms that TCDD is not as toxic as has been thought, the ramifications of such a finding will be interesting even though frustrating. Rather than being an evil spirit or terribly toxic poison, dioxin would have been merely a minor impurity. Thousands of Vietnam veterans could have been spared worries about possible exposure to herbicides. Foresters and farmers could still be using very effective formulations. Many Italian women should not have had abortions. Environmentalists could devote themselves to activities that are really deserving of support. Let us hope that we will be smarter in the future."

Vernon A. Stenger

9-22-89